RDS-STUDENT
Software for Aircraft Design, Sizing, and Performance

Daniel P. Raymer
President, Conceptual Research Corporation
Playa del Rey, California

Developed for use with
Aircraft Design: A Conceptual Approach
Fourth Edition

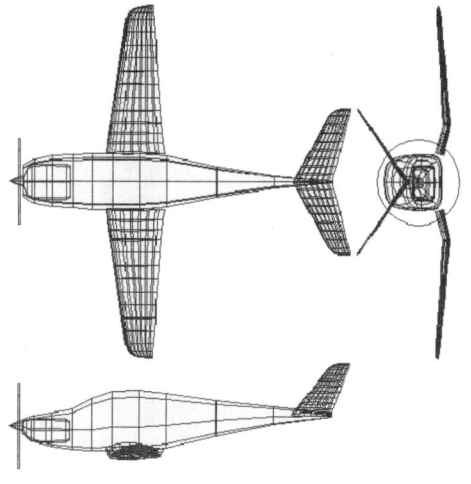

EDUCATION SERIES

Published by the American Institute of Aeronautics and Astronautics, Inc.
1801 Alexander Bell Drive, Reston, Virginia 20191

American Institute of Aeronautics and Astronautics, Inc., Reston, Virginia

ISBN 1-56347-831-5

Copyright © 2006 by Daniel P. Raymer. Published by the American Institute of Aeronautics and Astronautics, Inc., with permission. All rights reserved. Printed in the United States of America. No part of this publication may be reproduced, distributed, or transmitted, in any forms or by any means, or stored in a database retrieval system, without prior written permission of Daniel P. Raymer.

RDS-Student User's Manual

TABLE OF CONTENTS

<u>INTRODUCTION</u>   1

<u>RDS OVERVIEW</u>   3

<u>RDS OPERATION NOTES</u>   7

PROGRAM START   7
USING RDS WITH WINDOWS™   7
STARTUP PROBLEMS   8
FILENAMES   9
DATA/MENU INTERACTION   10
NUMBER ENTRY   11
UNITS   12
GRAPHS AND GRAPH EDITING   12
VIEWING TEXT FILES   15
DOS SHELL   15
RDS LIMITATIONS   15
RDS-EZ   17

DESIGN LAYOUT MODULE 18

OVERVIEW 18
RDS-DLM OPERATION 20
RDS-DLM COMPONENT GEOMETRY AND AXIS SYSTEM 21
RDS-DLM MAIN MENU 26
VIEW MENU 27
COMPONENT EDIT 30
ASSEMBLE COMPONENTS 34
UNDO LAST CHANGE 34
FILES & ANALYSIS 35
MENULESS VIEWING 36
DLM MAIN MENU OPTIONS 38

FUNCTIONAL ANALYSIS MODULES 39

FUNCTIONAL ANALYSIS OVERVIEW 39
AERODYNAMICS MODULE 40
WEIGHTS MODULE 47
PROPULSION MODULE 50

AIRCRAFT DATA FILE MODULE 55

CAPABILITIES ANALYSIS MODULES 64

CAPABILITIES ANALYSIS OVERVIEW 64
SIZING & MISSION ANALYSIS MODULE 64
PERFORMANCE ANALYSIS MODULE 69
COST ANALYSIS MODULE 71

INTRODUCTION

RDS-Student is a sophisticated yet friendly aircraft design and analysis computer program developed for the conceptual design and analysis of new and derivative aircraft. RDS-Student was written primarily as an instructional aid and complement to the bestselling AIAA textbook *Aircraft Design: A Conceptual Approach* by Dr. Daniel P. Raymer, and is suitable for use by students of aircraft design for learning how to do conceptual design studies and preliminary performance predictions.

RDS-Student is especially useful for automating the tedious calculations required in a typical "Senior Design" class, allowing the students to spend more time experiencing the process and flow of aircraft design. An expanded, full-featured version of this program called RDS-Professional is available for use by aircraft designers in industry, government, and academia.[*]

This latest version of RDS features an all-new Design Layout Module (CAD) with greatly expanded capabilities for design layout. This new DLM is the result of years of development and allows the rapid development of a new or derivative design layout. It has automated routines for quickly creating and modifying most of the components used in aircraft design including wings, tails, fuselages, landing gear, engines, seats and more. It uniquely "decouples" the screen display from the geometry commands allowing certain design tasks to be done in any desired view using the aircraft coordinate system, even when mouse inputs are being used. This is described in detail below.

RDS-Student uses the design and analysis methods of Dr. Raymer's bestselling textbook, distilled from the classical and time-proven first-order techniques commonly used in industry design groups for early visibility into design drivers and options. These methods include

---

[*] The license agreement for RDS-Student specifically forbids its use for professional or commercial applications. For such applications please contact the author (draymer@aircraftdesign.com) for information about RDS-Professional. RDS-Student may not be installed on a network or multi-user computer (computer lab). For such installations please contact AIAA for site license information.

aerodynamics, weights, propulsion, and cost, as well as aircraft sizing, mission analysis, and performance analysis including takeoff, landing, rate of climb, Ps, fs, turn rate, and acceleration. These are automated in user-friendly modules that permit a tremendous quantity of calculation including trade studies and optimization in the early stages of design. RDS provides graphical output for drag polars, L/D ratio, thrust curves, flight envelope, range parameter, and more.

RDS-Student does not provide "automagic" design optimization, as that would lessen the learning experience. In RDS-Student, trade studies and optimization are done by explicitly changing the trade parameter (T/W, for example) in the analysis input modules, then rerunning the sizing and performance modules. RDS makes this easy and quick.

Dr. Raymer, author of RDS, is keenly interested in continuously improving RDS and in fixing any bugs that may have crept in. Kindly send an email to *rds@aircraftdesign.com* if you have any suggestions or have found a bug – be as specific as possible to help in its eradication. Also, this release of RDS includes the all-new Design Layout Module as detailed below. While a substantial effort has been made to "bug-proof" it, there may be some anyway as in any new release. Free updates will be posted as needed at *www.aircraftdesign.com/rds.html,* along with RDS information including Tips & Tricks and RDS sample data files.

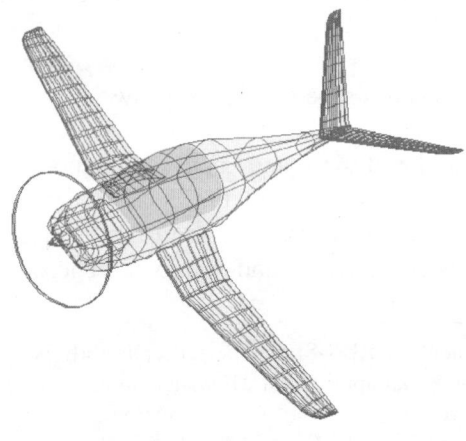

RDS OVERVIEW

RDS-Student is an aircraft conceptual design and analysis computer program written to provide design students with a fast and user-friendly tool for the design of new aircraft concepts. RDS-Student includes a 3-D CAD module for aircraft layout and a full set of analysis tools leading to performance and sizing calculations. RDS consists of several compiled execute files plus associated data files. RDS will run on any newer IBM-PC compatible system, and is normally run in Windows^{TM} using a pre-defined shortcut icon. RDS is structured to work equally well with mouse operation or by using arrows and other keys – it is up to the user. Commands are selected from "pick-and-click" menus, with defaults generally being the option most likely to be selected by the user.

Figure One illustrates the organization of the various RDS program modules. The Design Layout Module allows you to develop a design layout using a wide variety of commands and capabilities including direct creation of common components such as wings and engines, and section-by-section creation of complicated shapes when desired. An enhanced surface lofting method called "SuperConics" is employed, offering the ease of use of traditional conics but allowing more complicated geometry without the need to piece together several conics. This is detailed in the DLM description below.

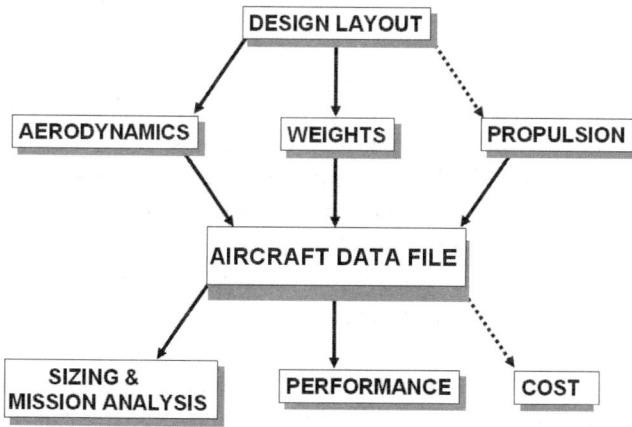

Figure One. RDS-Student Organization

Once the design is complete, a Geometric Analysis routine extracts from the CAD file the information required for Aerodynamics and Weights analysis and loads that data into those modules. Note that it is not necessary to use the RDS Design Layout Module to make use of the analysis modules of RDS. Instead, you can provide the required geometric data from a paper drawing or from another CAD system by typing the data into the appropriate locations in the RDS analysis module input matrices. While this sounds laborious, it actually takes less than an hour to manually input all required data should you choose to go this route.

The Aerodynamics Module estimates parasite drag (subsonic and supersonic), drag due to lift, lift curve slope, and maximum lift, based upon classical techniques. RDS also includes a longitudinal stability and control capability that estimates static margin and creates a trim plot, and calculates and stores a trim drag adjustment.

Weights and balance are estimated statistically from inputs defined in a spreadsheet-like matrix, after selection of aircraft type (fighter, transport/bomber, or general aviation). Results are presented in standard weight report format including structures group weight (fuselage, wing, etc.), propulsion group, equipment group, and useful load group. Center of gravity is also determined.

Propulsion analysis includes installation analysis of thrust and specific fuel consumption for jet engines, and propeller thrust and specific fuel consumption for piston-prop and turboprop engines.

One feature of the RDS program gives great flexibility for analysis and also allows (almost forces) the user to check these analysis results before committing them to vehicle sizing and performance. Analysis results from the Aerodynamics, Weights, and Propulsion modules are *not* directly used by the Sizing & Mission Analysis Module or the Performance Module. Instead, these results are stored as data sets in the Aircraft Data File, which is then used for sizing and performance.

Thus, if you change the inputs in, say, the Aerodynamics Module there is *no* direct effect on sizing or performance unless you first run the analysis in the Aerodynamics Module and instruct RDS to store the results in the Aircraft Data File. Some users have suggested re-coding RDS so that any change in any module would automatically "flow down" to all other modules, but this has not been done because it would eliminate much of the flexibility of RDS and its abilities to do trade studies. YOU are in command, and YOU decide when and how the data actually used for sizing and performance is to be revised.

The Aircraft Data File Module, shown in the middle of Figure One, is the analytical heart of RDS. It allows you to review, revise, graph, and if desired, directly enter the data that defines the aircraft for sizing, range, and performance calculations. The Aircraft Data File includes calculated C_{D0}, K, C_{L-max}, installed engine thrust and specific fuel consumption, and other sizing inputs such as the As-Drawn Takeoff Weight and the Payload Weight. Parameters are stored in matrix format as a function of one or two variables (such as velocity and altitude). Note that sizing, mission analysis, and performance analysis *cannot* be run if the Aircraft Data File is missing or incomplete.

Data for the Aircraft Data File can be calculated elsewhere and directly typed in, thus skipping the modules shown above it in Figure One. Normally, though, the Aerodynamics, Weights, and Propulsion Modules will be used to analyze a design and load the appropriate data into the Aircraft Data File. Once the Aircraft Data File is properly created, by whatever means the user prefers, its data can be used in the Sizing & Mission Analysis Module and in the Performance Module.

In the Sizing & Mission Analysis Module, a mission is interactively created by selecting mission element types (takeoff, cruise, combat, etc...) then entering the required data in a matrix format. Then the aircraft can be sized to that mission, resulting in the sized design takeoff weight and the fuel weight to perform that mission. The aircraft may be sized assuming either a rubber-engine or a fixed-size-engine, selected from the Options menu in the Sizing & Mission Analysis Module.

It is also possible to analyze the as-drawn aircraft for range and loiter capabilities over a defined mission, maintaining the as-drawn takeoff gross weight (i.e., not resizing to meet the specified range but instead varying the range or loiter until the as-drawn aircraft can make the mission). This is used for analysis of an existing aircraft or to determine alternate mission capabilities of a conceptual design that has been previously sized to another mission.

The RDS-Student Sizing & Mission Analysis Module includes a number of commonly used sizing trades. These calculate the effect on aircraft gross and empty weight of parametric variations in parasite drag, drag due to lift, specific fuel consumption, dead weight, payload weight, and range. Also, the range-payload tradeoff can be determined. Results of these trades are graphed automatically. Cost can be calculated and used as the trade measure-of-merit. This is a key difference between RDS-Pro and RDS-Student: students are required to do the trade studies themselves by changing the parameters in RDS, rerunning sizing and performance, and graphing the results themselves as a learning experience. Professionals simply want the results, right now!

The Performance Analysis Module uses the data in the Aircraft Data File to calculate performance such as takeoff, landing, rate of climb, Ps, fs, turn rate, and acceleration. This module will also graph the aircraft flight envelope, rate of climb, Ps contours, fs contours, and specific range parameter (nmi/lb). Most performance calculations are simply selected from an initial menu, but the takeoff, landing, individual Ps, and acceleration calculations require defining the input values in a Performance Analysis file that can then be stored and retrieved for future analysis.

The Cost Module uses the statistical DAPCA IV model for development and procurement costs. Life Cycle Cost and Airline Economic costs are calculated by yearly accumulation of costs based upon user inputs, adjusted for the inflation rate.

RDS OPERATION NOTES

PROGRAM START

To install RDS, use the utility INSTALL.EXE that is on your CD-ROM. Run it, and then follow instructions. Or, you can do the installation yourself. Create a new directory for RDS (/RDS is preferred), and copy the entire contents of the RDS CD-ROM into the RDS directory. Do not use a directory name that is more than eight characters or that has non-alphanumeric characters.

If running under Windows™, set up an RDS Shortcut as described in the next section. Then click on the RDS Icon, or enter RDS from the DOS prompt, and you are up and running!

Note that if you copy the RDS program files to another directory you may create problems with the RDSCONFG file, which records your preferred program options. If you get error messages, first try deleting RDSCONFG and rerunning RDS.

USING RDS WITH WINDOWS™

RDS is, at its core, a DOS program. However, these days virtually all RDS users are running it under Windows™. To set up a shortcut Icon to start RDS from most versions of WINDOWS, right-click on the desktop and select NEW-SHORTCUT. Enter the command line (probably RDS/RDS.EXE) and the program name (RDS). Then right-click on the new RDS shortcut icon and select PROPERTIES. Under PROGRAM, select Run-Maximize and Close on Exit. Under Memory, everything should be AUTO except EMS which should be NONE. Under Screen, it should be Full Screen and Default, and everything else checked. Under MISC, everything should be checked except Screen Saver, Suspend, and Quick Edit.

For instructions to run RDS under earlier versions of Windows, or if you encounter any problems, please visit *www.aircraftdesign.com*.

Under earlier versions of Windows, one could copy an RDS screen image into the Clipboard by pressing Alt-PrintScrn, then directly paste it into another program such as Powerpoint™ or Excel™. This is not possible under newer version such as Windows XP, so this latest version of RDS has its own Bitmap capture routine for graphics, and text file output routines for the rest. Bitmap images can be modified using the Windows Paint™ program or a more-capable commercial graphics program to invert colors, crop, shade, and carry out other "artsy" improvements to the RDS line art. Also, it is a good idea to reduce the image size by converting to a compressed file format such as (.JPG), (.GIF), or (.TIF) when saving the image file.

Text information can be passed to Windows applications by selecting Print to File for inputs or outputs, then opening the file in your application program. When passing tables of results such as mission sizing, select the Courier font because it maintains equal spacing between characters and thus maintains the alignment of RDS text columns.

STARTUP PROBLEMS

If RDS, or especially the Design Layout Module, will not run, you may not have sufficient free DOS memory. Check your free memory by entering MEM from the Command (DOS) prompt under Windows *Accessories*. You need about 580k-630k free memory to run RDS. You can free up memory using the DEVICE=HIMEM.SYS driver, and by setting memory as shown in the sample CONFIG.SYS file available at *www.aircraftdesign.com*. However, this should not be a problem for any modern PC unless its own setup restricts memory access for DOS programs.

When starting RDS for the first time, you will be prompted to identify the directory in which you have stored your copies of the RDS program files. You will also be asked for your preferred units (FPS or MKS). RDS will store this information in file RDSCONFG for future use, along with any customization of RDS which you perform, such as changing the mouse sensitivity parameters and desired graphing options. These are updated every time RDS is used on your computer. If this file RDSCONFG becomes corrupted or modified in some way, you may experience erratic operation such as ERROR 76, PATH NOT FOUND. To fix this or other operational problems, try deleting the RDSCONFG file and restarting RDS.

FILENAMES

RDS uses specific filename extensions beginning with D (Data) to identify and access data from the various modules. Upon entering RDS you are prompted to give a Project Filename (maximum eight characters), which, with the appropriate extension, becomes the default filename for the various modules' data. For example, if your project filename is MYPLANE, then your default files are:

```
Design Layout Module:                    MYPLANE.DSN
Aerodynamics Module:                     MYPLANE.DAA
Weights Module:                          MYPLANE.DWT
Propulsion Module:                       MYPLANE.DPR
AIRCRAFT  DATA FILE Module:              MYPLANE.DAT
  (Uninstalled jet engine data)          MYPLANE.ENG
  (Uninstalled turboprop engine)         MYPLANE.TBP
Sizing & Mission Analysis Module:        MYPLANE.DMS
Performance Analysis Module:             MYPLANE.DPA
Cost Module:                             MYPLANE.DCA
```

When RDS prompts for a filename either to read-from or to save-to, it will offer your current project filename plus the appropriate extension as the default. You may instead choose to enter another filename, upon which RDS will list the available filenames of the appropriate extension.

For example, if you are using project filename NEWJET, then your baseline MISSION SIZING FILE would be called NEWJET.DMS, using the default filename. You could also store an alternative mission, which you could call NEWJET2.DMS, or MISSION2.DMS, or whatever you wish. When you next enter the Sizing & Mission Analysis Module, RDS would offer NEWJET.DMS as the default file, but you could instead get the alternative mission file. RDS will list all available files with the .DMS extension, and ask you to input the desired filename. You must then spell out the entire filename - including the extension!

You can use filename extensions other than those listed above, but be aware that RDS will not list files with other extensions when prompting you to input a filename. Thus you could call your alternative MISSION SIZING FILE something like MISSION.WOW, but RDS won't know to list that as an available Mission Sizing file so you would have to input the filename from memory (yours, not the computer's). Therefore, it is strongly recommended that you stick to the same extensions that RDS uses (.DMS in this case). Note that the jet and turboprop engine data files for manufacturer's uninstalled thrust and SFC must have extensions .ENG or .TBP, or RDS won't recognize them as such.

Also, beware of forgetting to give an extension when typing in a non-default filename, especially when saving a file. RDS will save the file under a new, no-extension filename. When you next try getting the edited file (now remembering to use the extension!) you will instead get the old, unedited version of the file.

RDS does check before reading a file to insure that it is in fact the correct type of file, ie, that you haven't told it to read a Performance Analysis file into the Sizing & Mission Analysis Module. To do this, RDS internally stores and tests on an integer data type code.

To change the current project filename, select the New Project Name option from the main Options menu.

DATA/MENU INTERACTION

RDS menu selection and data box motion is via mouse or arrows. Menus initially have one option highlighted, which is the default option. Use the mouse or arrows to select the desired option, then press Enter or the left mouse button. Page Up, Page Down, Home, and End are also available to quickly jump around matrices and menus. The Escape button will generally exit a menu or matrix, taking you "up" a level, although it is preferable to use the provided option to exit a menu. Yes/No prompts can be selected with mouse, arrows, or by entering Y or N.

The selected item for menus in all modules except Design Layout is displayed as a brighter color, or it can be displayed as a reverse-shaded box for greater visibility. If the cursor is not sufficiently visible on your computer when a menu is displayed, press C to toggle between the color and the reverse-shaded cursor. In the Design Layout Module, the selected item is shown in red with a ">" preceding it.

Much of the RDS data input and manipulation occurs in spreadsheet-like data matrix displays. To select from the command menu at the bottom of a data matrix display, simply run the cursor off the bottom of the data matrix to the menu area. Then select the desired menu option and press Enter, or the left mouse button. Note that Page Down is especially useful for jumping directly into the command menu at the bottom of a matrix.

PRINT/GRAPH provides a menu of options including printing the current input array directly to your printer (which must be on-line) or to a text file, and creating various graphs depending upon which module you are in. Note that many computers these days do not even have a printer directly attached to a printer port, instead relying upon Windows to handle all printing tasks. If so, you should always select Print to File, then use a Windows program such as Word or Powerpoint to format and print the desired output.

OPTIONS brings up a menu of options appropriate to the RDS module in use, such as changing the origin of graphs or the mouse sensitivity parameters.

EXIT/SAVE exits the matrix, giving you a number of choices as to whether to save and where to go next. Other menu options are described in the module write-ups.

NUMBER ENTRY

RDS number entry in the data matrices is via the numeric keypad, with number sign entry following the digit entry as with a ten-key adding machine. In other words, to enter the negative number "-25" you should enter "25" followed by - . No [Enter] is required. This ten-key entry method speeds up entry of a large amount of data. However, if you desire you may also enter numbers the regular way followed by [Enter], ie, "-25"

[Enter]. This does cause a redraw of the entire screen, which slows down operation.

UNITS

RDS is fully capable in both Metric (MKS) units and in units of the British Imperial (FPS) type, now used primarily in the USA. In the FPS units RDS uses thrust in lbs., C in 1/hr, altitude in feet, and velocity in either knots or Mach Number. In the MKS units RDS uses thrust in kN, C in mg/Ns, altitude in meters, and velocity in either km/hr or Mach Number. It is not necessary to identify whether knots (or km/hr) or Mach Number are used because RDS assumes that any velocity values less than 10 were given in Mach Number.

Units are selected in the various Options menus, or they can be toggled from one to the other by entering # (the "pound" sign, to help remember) from any data matrix or menu. You can jump from one to the other at will, and enter or view all data in whatever units you prefer. For example, even metric users often want to enter velocities and ranges in nautical miles and knots because that's how the pilots talk. This is easy - just toggle to FPS, enter the value, and toggle back to MKS.

There is also a units conversion utility in the Options menu allowing you to type in a value in one system and have it converted to the other system. This includes typical aircraft design-related values such as wing loading and specific fuel consumption.

GRAPHS AND GRAPH EDITING

RDS will graph inputs, analysis and performance results, and aircraft data such as engine thrust and drag polars. Graphs are plotted with a sophisticated curve fit routine and the graphing module includes editing capabilities to create viewgraphs and report graphs. Finished RDS graphs can be "captured" as bitmap images for use by Windows application programs for rapid production of technical reports and briefing charts.

When a graph has been displayed on the screen, a number of options are available. To maximize resolution and clarity, the entire screen is filled with the graph. Options can be displayed by pressing H (help), bringing up the box shown below. The desired option is selected by pressing the appropriate letter or symbol on the keyboard (it is not necessary to press H first to use these options).

```
GRAPH Options:                          B = BITMAP
A = NEW AXIS          Z = ZOOM          C = CHANGE COLORS
| = ADD/REMOVE GRID
E = Edit  <Box Line Circle Move Erase Text Done>
```

A screen image bitmap can be created simply by pressing B. Each bitmap created is given a name derived from the design file name in use, truncated and with a number added. You should open such bitmaps in a graphics program such as Photoshop, IrfanView, or MS-Paint, then revise or invert the colors if needed.

The graph axis may be changed by entering the letter A (Axis). This erases the graph, including any editing or lettering as described below, and prompts for a new XMIN, XMAX, YMIN, and YMAX. These are entered separated by commas. For example, if the rate of climb graph includes velocities up to 500 kts and altitudes to 20,000 feet for your 200 kt homebuilt design, you could re-graph only the data range of interest by pressing the letter A then entering, say; 0,200,0,10000 [Enter]. You may zoom the graph to fill the grid by pressing Z.

Press C repeatedly to change the graph colors. Pressing "|" (it looks like a grid line) will remove or restore the grid lines internal to the graph.

Graphics editing is entered by pressing E when a graph is on the screen. When E is pressed, a cursor (+) appears and a list of single-key options appears to the right of the graph, as follows.

```
L - Line
B - Box
C - Circle
T - Text

S - Slow    (slow cursor speed when arrows pressed)
D - Default (medium cursor speed)
F - Fast    (fast cursor speed)

E - Erase
M - Move
Q - Quit
```

Line, Box, and Circle all require two inputs. The first is the start point of a line, one corner of a box, or the origin of a circle. The second point is the end of a line, the other corner of a box, or a point on the circle defining the radius. Note that for a box, entering the plus symbol (+) as the second input causes the box to be filled. For any of these, pressing Escape will cancel it.

Text permits entry of text at the current cursor location. Press Enter when done. The backspace and Escape keys also function. Note that if you try to input more text than room permits, the display will wrap to the next line and probably mess up your graph! Also, you are limited to 30 keystrokes including backup and re-enter if you make a mistake.

The speed and sensitivity of the cursor movement can be changed by pressing S for Slow, D for Default (medium), or F for Fast.

Erase and Move both require two inputs, namely the upper left and lower right corners defining a box to be erased or moved. For Move, after the box is defined you move it with arrows or the mouse then press the space bar when you have aligned the box as desired. Note that the entire display within the defined box is erased or moved, including the graph axis, labels, etc...

Press Q when finished with graph editing. RDS will erase the cursor and edit options list, permitting you to copy your display.

Various graphing options can be selected from the Options menu of the various RDS modules prior to creating the graph. These parameters include the desired default graph origin and the use of a curve fit or straight-line interpolation, and are saved in file RDSCONFG when you exit RDS.

To exit a graph, press Return or click the mouse button. Either input will erase the graph and return to the appropriate RDS menu or matrix.

VIEWING TEXT FILES

RDS permits viewing text files such as those created during aerodynamic or sizing analysis. From any menu or data matrix, press (shift) < and give the filename (if you have just saved text output RDS will automatically view that file). Use the mouse, arrows, and page up/down to scroll through the file.

DOS SHELL

RDS permits executing a DOS command or program without exiting RDS or losing your working files. From any menu or data matrix (except in DLM), press SHIFT > then enter the DOS program name, or ENTER to go to the DOS prompt (which is ">"). Return to RDS by exiting your other program or typing EXIT at the DOS prompt.

RDS LIMITATIONS

RDS employs certain assumptions that may cause erratic operation if the user attempts to use RDS for unusual designs. Velocity is assumed to be Mach number if values given are less than 10. Thus, one cannot use RDS for an aircraft flying at under 10 kts, and, for other reasons, over about Mach 5.

Another limitation in RDS is intended to protect the user from obtaining false extrapolations of data. RDS will only calculate performance within the provided range of the engine thrust data in the Aircraft Data File. For example, if the engine thrust data only goes up to Mach .8, the flight envelope will be cut off at that point even if there is still excess thrust-minus-drag. To ensure that the indicated maximum speed on the flight envelope really is the true maximum, you must provide thrust data at higher speeds. You can, of course, check if thrust equals drag at the indicated maximum speed by plotting the 1-g Ps contours.

Also, it is a good idea to provide thrust and SFC data for the entire matrix of altitudes and velocities, even though the engine may not be rated for, say, high speed/low altitude operation. If zero values are left in the matrix, RDS may linearly interpolate from real values to zero values, creating strange looking diagonal lines on the graphs, and possibly creating errors in the sizing and performance results.

Another thrust limitation is that RDS tests on {T/W > 2} as an indication that instead of T/W and W/S, the user has provided thrust (T) in pounds and wing area (S) in square feet. Therefore, an aircraft with thrust in excess of twice the takeoff weight is not permitted!

The standard atmosphere model in RDS cuts off at 150,000 feet. The highest turning load factor RDS calculates is 9 g's. Extreme negative values of Ps, fs, and rate of climb are clipped off the graphs so that the more-relevant values can be read. If a supersonic design has a negative Ps (thrust-minus-drag) in the transonic regime but a positive Ps at higher speeds, the logic to graph the flight envelope may be confused. In such a case, the 1- g Ps contours can be used to manually construct the correct flight envelope showing the "bubble" of sustained supersonic flight.

To speed up execution, RDS is compiled with the BREAK button disabled. If for some reason RDS should hang up completely, you must reboot. This happens rarely, and is usually due to incomplete or impossible inputs in the Sizing & Mission Analysis Module.

RDS-EZ

The main menu of RDS offers something called "RDS-EZ." This is simplified method of developing the inputs required for range and performance analysis, assuming an aircraft with a normal design arrangement. It works as if a design expert were to ask you some overall questions about the design and then approximate the inputs required to estimate range and performance.

RDS-EZ asks you for about 20 geometric and weights numbers defining your design, such as length, wing area, and whether the landing gear can retract. Then it automatically develops the regular RDS input files for aerodynamics, weights, and propulsion analysis. The program allows you to override these analysis results in several areas such as empty weight, in case you know more about the design than RDS-EZ can estimate.

After RDS-EZ creates these input files, you should go view them, change them as desired, and run the analysis.

RDS-EZ is just a quick approximation of the inputs needed for analysis. You can use RDS-EZ to get started on developing analysis files or to get a 5-minute answer, but you should not rely on the results from RDS-EZ for anything more.

DESIGN LAYOUT MODULE

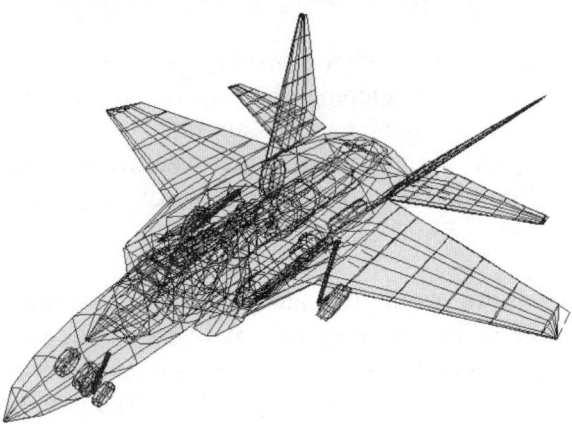

OVERVIEW

The Design Layout Module in this version of RDS is a completely new and original program for computer-aided aircraft conceptual design.[*] The new RDS-DLM permits rapid aircraft configuration layout using mouse-driven interactive computer graphics, allowing the designer to develop a new concept or modify the previous baseline design in a methodology custom-tailored for the advanced aircraft design environment. The design capabilities of this module include wings, tails, fuselages, nacelles, seats, canopies, and other required components. RDS-DLM allows interactive assembly of the aircraft[†] using top, front, or side view, or even in an isometric, orthographic, or perspective view.

---

[*] While the previous design module had some good features and was used for years by this author and others, it was not fully integrated with the rest of RDS and its geometric representation was not suitable for other than the earliest stages of design. Given that it was programmed in an obsolete language, it made more sense to start over.

[†] Throughout the program and documentation the word "aircraft" is used. Don't let this limit you. RDS-DLM also works well for airships, launch vehicles, UAVs, and even ships, submarines, and other vehicles.

While perhaps not as graphically powerful as a commercial CAD program such as Catia™ or Unigraphics™, RDS is uniquely suited to Aircraft Conceptual Design because it knows what an airplane is. RDS-DLM has dozens of airplane-specific design capabilities to rapidly create, modify, and analyze your design. RDS-DLM even uses the aircraft industry SAWE RP8A (Mil-Std-1374a) Group Weight Statement component categories to identify the nature of the design's various parts and to simplify the analysis interface.

When finished with a design, RDS-DLM provides a simple data interface to the Aerodynamics and Weights Modules and hence to all RDS performance analysis and optimization. It literally takes under a minute to analyze the design geometry and make it available to the analysis routines. RDS-DLM also produces the tabular geometric information required for reports and further analysis.

RDS-DLM features a user interface that is mouse-driven but not exclusively cursor-driven. Many design changes can be done by "flying" the selected geometry with the mouse or keyboard instead of traditional "click & drag." These design operations can be done in any screen view - the user inputs are in the aircraft axis system, not in screen coordinates. Design is therefore more three-dimensional and intuitive, once you get used to it!

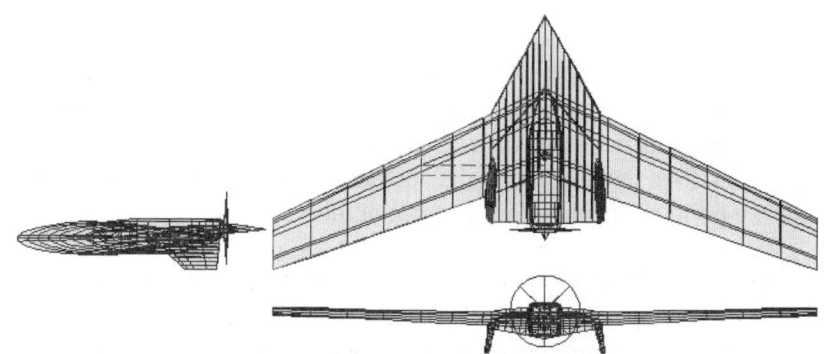

Boeing Educational Project UAV designed in RDS
(D. Raymer & R. Dellacamera, Instructor/Design Consultants)

RDS-DLM OPERATION

The Design Layout Module, like all RDS modules, is selected from the RDS main menu. RDS-DLM provides menus for all design and viewing operations, ideal for beginners and occasional users. In these menus the highlighted item is "hot" – click the left mouse button or press Enter and that item is selected. To navigate up and down the menu, use the mouse or up/down arrows. PageUp and PageDown also work. If a menu is longer than the screen height allows, just run the highlighted selection off the page and the menu will jump to the next page automatically. For such a long menu, Home and End can be used as well. Escape (ESC) can be used to exit a menu or option, but it is preferable to use the menu exit or cancel option because the ESC button may return you to an unexpected part of the program in some cases.

Yes-No prompts can be entered as "Y" or "N" (not case sensitive), or can be selected by left-right mouse or arrow inputs followed by a mouse button click or Enter.

A powerful capability for DLM experts is Menuless Operation, which removes the menus to permit full use of screen space, then allows viewing and component movement and rotation to be done using a variety of mouse, arrow, and Hot Button inputs. Menuless Operation is described in a separate section below.

A mentioned above, RDS-DLM does not use "click & drag." Cursor input such as movement of points is always done with two distinct inputs - first the point is identified, and then its desired location is given. This allows toggling to fine movement of the mouse or use of numerical inputs of either the point coordinates or the point number.

In many ways RDS-DLM is more like the RDS analysis graphs than the text-mode data entry screens. RDS-DLM uses the same color schemes as the graphs, and a similar cursor implementation, but adds menus based on those in the analysis modules. In any case, a user who is familiar with the look and feel of the RDS analysis modules will feel right at home in the new RDS-DLM.

RDS-DLM COMPONENT GEOMETRY AND AXIS SYSTEM

RDS-Student components can be defined in one of two geometric schemes, both using cross sections stacked in the component local X-direction. In the simpler scheme, those cross sections are formed from stored points which are the actual surface definition. This scheme is usually used for wings and tails in which case the stored points are the actual airfoil coordinates.

For most components the cross sections are defined by one or more "SuperConic" curves. SuperConics are modified 4th degree Bezier polynomials with a quartic mathematical representation[*] which allows their shapes to be designed and controlled in a manner similar to the classic conic lofting used on aircraft since the 1940's when they were first used for the North American Aviation P-51 Mustang.

SuperConics are each defined by five control points. The curve has two endpoints plus a point somewhere along the tangent line from each of the endpoints. The fifth point is called the "shoulder point" and is on the curve, somewhere in its middle.[†]

Design using SuperConics[‡] is very similar to design using Conics as in the previous RDS Design Layout Module. The figures below show the two types of curves – both have start points (A and B) and a shoulder point (S). The main difference is in the control of the tangent lines to the end points, shown as dotted lines below. For the classic conic these tangents

---

[*] Maier, Robert., "Quartic Curves," Rockwell/North American Aviation TFD-78-718 (unproprietary, unclassified, and unfortunately unavailable internal document), Los Angeles, CA 1978

[†] This is different from a classic Bezier 4th Degree Polynomial where the middle point would be floating off the line and acting like a "magnet" pulling the line to itself. Mathematically placing the middle point *on* the line makes it much easier to define and shape the desired curve.

[‡] The correct terminology for this curve form is "modified parametric 4th Degree Bezier Polynomial." The name "SuperConic" is this author's heartfelt attempt to make these elegant and intuitive curves a bit less intimidating!

must meet at the point labeled C, and C itself is used as a control point for conics.

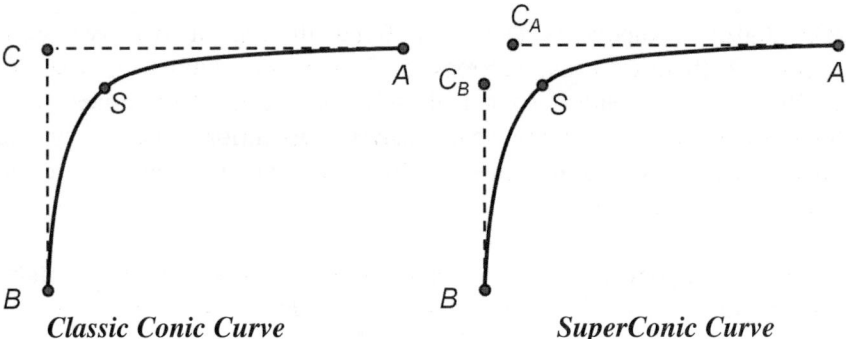

Classic Conic Curve   **SuperConic Curve**

For a SuperConic, there are two separate control points, C_A and C_B. They do not have to meet or even be close to each other. Also, with a 4th degree polynomial representation the SuperConic curve can include a reflex – unavailable in a regular conic. Curves such as those shown below are just as possible and easy to construct as the conic look-alike shown above.

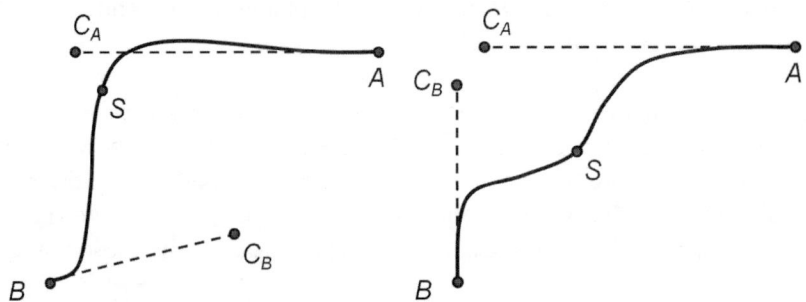

In RDS-DLM, connecting SuperConic curves share an endpoint so cross sections always have 5, 9, 13, 17, or 21 points corresponding to 1, 2, 3, 4, or 5 SuperConic curves. More section curves are possible, up to the RDS maximum of 100 points per section – but that would be silly. If you can't represent a cross section shape with just 2 or 3 SuperConics, you are probably doing something wrong!

For components defined by SuperConics, RDS stores these control points for each cross section and generates the actual points shown on the screen only when drawing. When reshaping a cross section, the SuperConic control points are shown and can be moved or otherwise modified, and the resulting section points are drawn.

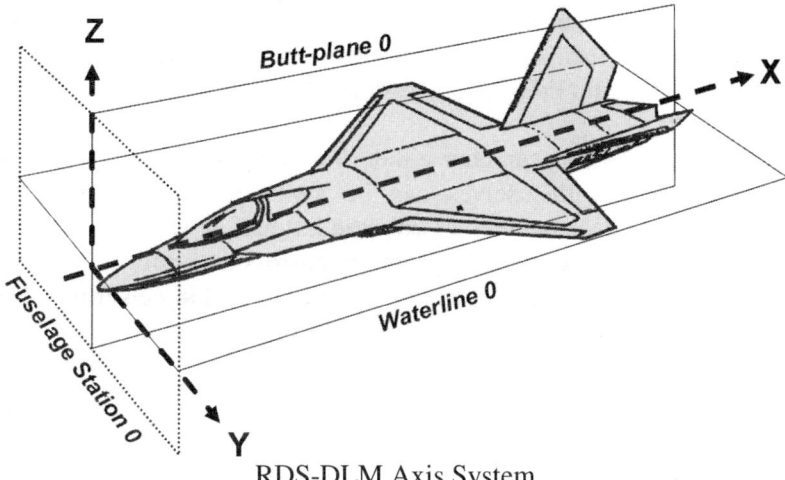

RDS-DLM Axis System

The RDS-DLM global axis system is shown above. Following classic drafting table practice at many companies, the X-axis points out the rear of the airplane, Z is up, and Y is out the left wing. X=0 does not have to be the nose of the aircraft but it often is, at least in the first layout. To meet the expectations of pilots and other airplane people, rotation directions are positive for right and up. Yes, this gives RDS-DLM a left-handed coordinate system! Don't worry, it doesn't affect anything and is easily revised when interfacing to other software where it may actually matter.

An RDS-DLM design consists of a number of components, each positioned in the aircraft by the location and orientation of its own local axis. Components can be positioned, moved, and rotated in several ways including Component Assembly, Component Edit-Parameter, and Menuless Move. Component location is defined by the origin of its axis system, normally at or near the front of bodies and at the root of wings and tails (and at the 0.25 MAC location). To display local axis systems see View Options.

Components such as wings, tails, and tires are usually "pre-rotated" so the local component X-axis is aligned with the global Y axis for wings, Z axis for tails, or -Z axis for ventral tails. Further component rotations are in addition to these. In other words, a Wing component with listed values for roll, pitch, and yaw all equal to zero actually has an initial yaw of 90 degrees, and the local X-axis will point out the global Y-axis. This is automatically set when such components are created, and can be reviewed and changed in Comp Edit-Parameters (see Axis Orientation).

RDS-DLM components can have either of two types of symmetry, or both types. Components themselves can be symmetrical about their own centerline, so the designer creates only the left half which is duplicated on the right half. The other type of symmetry mirrors the entire component to the other side of the aircraft, across the aircraft global centerline. These are illustrated below.

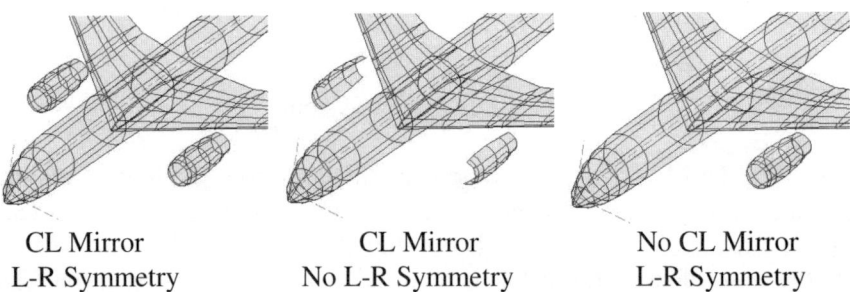

| CL Mirror | CL Mirror | No CL Mirror |
| L-R Symmetry | No L-R Symmetry | L-R Symmetry |

A final type of symmetry is only for wings and tails. These are normally mirrored across the aircraft centerline forming left and right surfaces. As an option for special design applications, they can instead be mirrored in the component X-direction across the component's root chord (local X=0).

The figure below illustrates this. In the left illustration we see normal mirror-image duplication of the tail across the aircraft centerline - the trapezoidal planform root chord is exactly at the centerline. This is typical for wings and tails, and is assumed in classical aerodynamic analysis including the methods in the RDS Aerodynamics module.

The center illustration shows a tail that is mirrored in the normal fashion but is moved off the aircraft centerline. This might be used for an aircraft like the F-15, where the fuselage is so wide that the effective tail area should not extend all the way to the aircraft centerline. When the tail is moved out in Y, a gap appears between the roots of the left tail and the right tail.

The right illustration shows the use of root-chord mirroring as might be used for an asymmetric aircraft such as Burt Rutan's Boomerang. The Y offset is the same as in the middle illustration, but the tail is mirrored across its root rather than across the aircraft centerline.

When using root-chord mirroring, yawing the wing or tail will create an oblique surface. However, the airfoils will no longer be defined in the freestream direction so you must work carefully to ensure that the analysis produced from such geometry is correctly modeled.

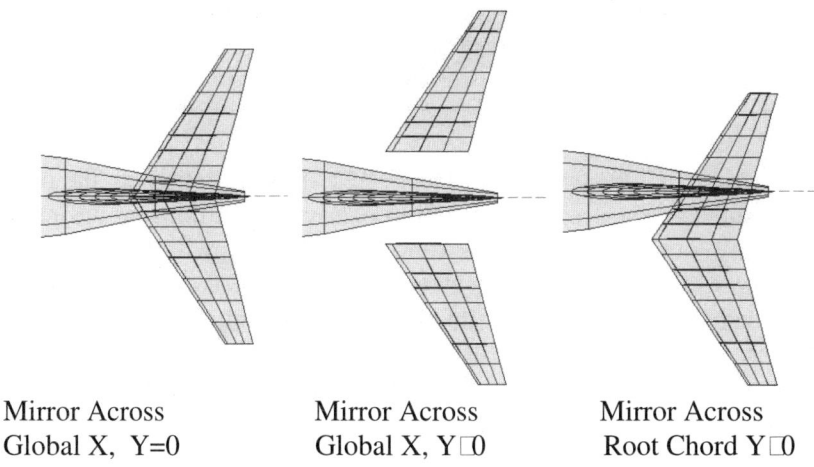

Mirror Across   Mirror Across   Mirror Across
Global X, Y=0   Global X, Y⊏0   Root Chord Y⊏0

RDS-DLM MAIN MENU

The Design Layout Module main menu has 9 options, each of which takes you to a submenu with a variety of further options and submenus.[*]

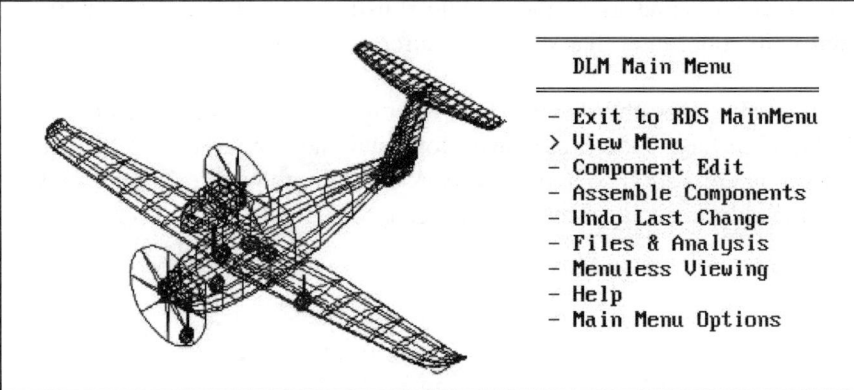

```
            DLM Main Menu
          - Exit to RDS MainMenu
        > View Menu
          - Component Edit
          - Assemble Components
          - Undo Last Change
          - Files & Analysis
          - Menuless Viewing
          - Help
          - Main Menu Options
```

The selection *Exit to RDS Main Menu* is like the exit options in the other RDS modules. First it offers the option of saving your current design. If you choose not to save at this time, your design will be kept in a temporary file and you will be given the option of continuing with its development should you return to RDS-DLM during that RDS session. However, when you close the RDS program completely, any work that was not saved will be lost. It is strongly recommended that you save your work when exiting RDS-DLM unless you are doing some sort of a temporary trade study and don't want the baseline design file to be overwritten. Such a trade study is the reason that RDS does not automatically save every change you make.

Selecting *Help* brings up an abbreviated version of this manual for the modern software user who cannot and will not read a paper manual! This includes the Menuless Operation procedures and Hot Buttons for easy reference (see below).

---

[*] If you start RDS-DLM and this menu is not shown, the last person who used RDS exited the program while Menuless Viewing was active – press Alt-M to restore the menu. RDS-DLM always restarts in the same mode and views as when it was last used.

View Menu

The View Menu allows the selection of a wide variety of views and view options. The first item on the View Menu (other than Exit) allows selection of components for view. RDS-DLM always brings the entire design file you have selected into the working temporary file, so a complicated design with dozens of components may create displays that are too cluttered for ease of use. Also, the time to draw depends upon the number of components, and there may be a noticeable delay when too many components are shown. The solution is simple – select only the components that you need to see for the current design activity.

The next three options bring up submenus to select various views. One peculiarity – if you are already looking at a perspective view and you select *orthographic*, you get a view from exactly the same direction but with perspective distance set to "infinity," i.e., no perspective distortion of the image. If you are already in an orthographic view including side, top, or rear view, and you select *orthographic*, the drawing will be exactly the same as before, until you rotate it. After all, a side view is just an orthographic view from the side! Select *isometric* if you want to see the design rotated to a view such as shown above.

A powerful feature is *Relative* viewing, offered in the Single View submenu. This requires that a component be selected, then the desired view is drawn "relative" to that component, i.e., in its axis system. For example, if the wing is selected for a Relative Side View display, the side view of the wing will be drawn. Since "side" means an X-Z view, the wing will be shown with its local X-axis to the right and its Z-axis up. In global terms this is an aircraft *front view*. Furthermore, the aircraft will be rotated to the wing's dihedral angle so that the left side of the wing is shown flat to the screen (see below). This allows making measurements and doing various design tasks with the rest of the aircraft visible and properly oriented – a powerful feature indeed. Component Shape commands begin by automatically showing such a relative view.

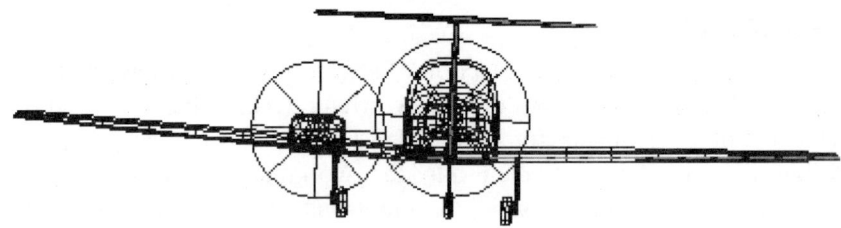
Relative side view of Wing

Move and *Rotate View* menus do just that. The *Speed* option changes the magnitude of the action provided by the offered options. Options for speed of move are *Nudge*, *Move*, and *Jump*, and relate to the amount of motion relative to the current screen scale. For rotations, *Nudge*, *Move*, and *Jump* provide angular changes of 1, 10, and 40 degrees respectively. These are reset to Move (or 10 degrees) every time you reenter these menus.

Zoom provides zoom options, including zoom to full screen and numerical input of zoom factor.

Measure and *Locate* are useful tools. *Measure* requires two distinct inputs – the starting point and the ending point. Click and Drag will not work, and when you release the button RDS will interpret it as a single entry at the cursor endpoint, and then patiently wait for the second input!

This two-input method is used to allow careful positioning of start and end points. The cursor can be moved by the mouse or by the use of arrow keys. Their sensitivity can be changed in the middle of cursor motion by pressing one of three speed buttons: *S* for *Slow*, *D* for *Default* (medium), and *F* for *Fast*. Note that these make a 3-button row on the keyboard for easy fingering, to use a musical term.

If the measurement is being made in a Side, Top, or Rear view, you can numerically enter the start or end points by pressing *N* (*Numeric*). This is useful to, for example, find the distance to some point on a component from the origin (enter 0,0 for the first point).

After the second point is entered, RDS prints the slantwise distance and the delta-X and delta-Y distances, along with the measured angle and its 90 degree reciprocal. Note that these distances and angles are in the plane of the screen, so make sure you are doing the measurement in the correct view orientation. Relative viewing as described above is very useful for this.

Locate is similar to *Measure*, but takes only a single input whose location is in either the global axis system or, if Relative viewing was used, in that component's axis system. *Locate* only works in a side, top, or rear view - in other views it would make no sense.

Measure or *Locate* will repeat until you exit. To exit from either, press Q or Esc.

Another View menu option creates a graphic bitmap (.BMP) of the screen image, stored in the currently-active directory. Each bitmap created is given a name derived from the design file name in use, truncated and with a number added. You should open such bitmaps in a graphics program such as Photoshop, IrfanView, or MS-Paint, then revise or invert the colors if needed. Next, save it with a more descriptive name and in a compressed graphic file format such as JPG or GIF. Bitmaps can also be created by pressing B during Menuless Operation or whenever a menu is on the screen.

There are many view options. You can choose to show the wing trapezoidal planform (always shown in dotted red), show it for all wing and tail surfaces, or not show it at all. To aid in positioning, trapezoidal planforms are shown with lines representing the .25 & .4 MAC positions. Normal aircraft will have their center of gravity somewhere between these points.

You can also show the axis systems, global and local, and show vertical lines at each section's centerline. You can toggle the mirroring of symmetric components and the mirroring of components across the global (aircraft) centerline.

The screen is normally cropped on the right side so that the menu does not cover up part of the aircraft display. You may wish to turn off the screen crop if you regularly use Menuless operation. The Crop Screen option is a toggle which is normally "on," so the menu option will usually show "Don't Crop Screen."

Display colors can be changed in View Options. This option is a toggle – each time you select it you get a different combination of colors. After you have toggled through all 10 options it starts over again.

Component Edit

Selecting the *Component Edit* option in the main menu brings up a submenu with a wide variety of tools for the initial creation and later modification of aircraft components. These include *Create Component, Get Comp from File, Select Comp for Edit, Shape Component, Scale Component, Copy Component, Delete Component,* and *Comp Parameters.* These are obvious in their function and operation, and will not be detailed here. Several are noteworthy:

Create Component brings up a menu with many ways to quickly make new components, such as a Fuselage, Wing, Tail, Wheel, Gear Shock Strut, Engine, Seat, and more. Generic shapes such as Box and Cylinder are also available. After selecting the desired option, RDS prompts you to enter a name then select the component type. Top-level categories of aircraft component type are offered then a detailed submenu prompts you to pick the specific component type. The numerical codes shown are based on the aircraft industry SAWE RP8A (Mil-Std-1374a) Group Weight Statement component categories[*], with three additional digits added by this author to further clarify component type.

These component type codes can be changed using Edit-Comp Parameters, and are used mostly to identify whether each component should be included in the aerodynamics or weights analysis, and which analysis method should be applied. For example, a component of type [031-000:Fuselage] will be included in the aerodynamics and weights calculations and treated as a fuselage. A component of type [031-009:Payload Bay] is identified as a part of the fuselage and can be so grouped in later detailed design efforts, but in the RDS analysis is automatically ignored. Note that this is fully correct in the aerodynamics calculations – the payload bay is inside and has no effect. However, in the weights calculations the statistical equations used by RDS do not have an

---

[*] Get the full specification including detailed component categories at the website of the Society of Allied Weight Engineers, *http://www.sawe.org/*

input based on payload bay existence and geometry, so the user (you!) must carefully consider how to modify the RDS results to account for it, probably by applying a fudge factor or adding an estimated amount to the We-Misc category in the input file.

Note that RDS-DLM will be unhappy if you fail to specify one and only one reference wing component. If you have two or more wings, only one can be the "reference.". If you do not have any wings, such as a lifting body or airship, you should define a "fake" reference wing component. Give it Sref=1 which will cause drag results to be expressed as drag area (D/q). When you are working on the Aerodynamics input file, give this fake component an exposed area (Aexp) of .001 so that the fake wing doesn't produce real drag! Note that you cannot give it Aexp=0 because RDS will assume that you simply forgot to provide this term, and will use Sref for Aexp.

For retractable landing gear, the down position copies of the components should use one of the available "AltPos" codes (alternate position) so they are not used twice in the analysis.

Continuing, RDS will prompt you to define the component symmetry as described above, then prompt you for the desired symmetry options, namely left-right symmetric in the component axis system, then mirrored or not across the global aircraft centerline. Finally, RDS will prompt you for the inputs required for the type of component you are building. After you have finished creating the new component, you will almost always want to modify it to exactly match your desires – see Component Shape and Scale, below.

One nice RDS-DLM feature helps in building wing components. Normally a wing is defined by its area (S), aspect ratio (A), and taper ratio. If you are trying to model an existing wing, you may know the span and chord lengths, not these parameters. RDS-DLM will determine the wing planform geometry from any combination of three inputs of S, A, taper ratio, span, root chord, or tip chord. These six parameters are offered – select one, then enter the desired value. There are a few silly combinations which RDS watches out for – you can't define a wing with only taper ratio, root chord, and tip chord!

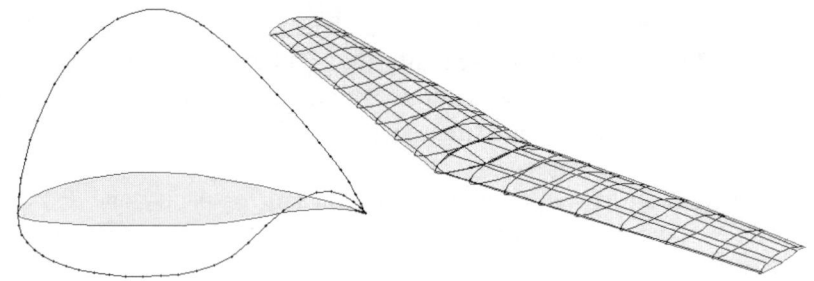

Selected Airfoil and Resulting Initial Trapezoidal Wing

After the trapezoidal planform parameters are input, RDS lists available airfoils and prompts you to pick one. The selected airfoil is drawn twice for your approval – once in actual proportions, and then a second time with the vertical axis stretched. The points are slowly plotted. Watch that they start at the rear, go forward over the top, wrap around the front, then go down the bottom back to the rear again. While you can do most design work with airfoils that are defined another way, certain RDS routines may be fooled and produce strange results.

If the component you need is vastly different from the available options, you can create it by making an "empty" component. Exit Create Component, go to Shape Component, select this new component, and make its cross sections. You can create a single cross section for the new component, copy it to other locations, then stretch, scale, and reshape those sections as desired. Or, you can make a new component the "hard way", one cross section at a time. You will probably use the numeric ("N") entry option a lot. However, this author usually prefers to start with some sort of simple shape such as a cylinder and modify it to the desired component shape rather than start with a truly empty component.

Once components have been created, various RDS-DLM design tools are available to rapidly change their geometry as the design evolves towards its final form. Component Scaling can be done in any axis or in combinations, including an option where length is stretched but cross section is reduced to maintain total volume (good for tanks).

One unique tool in the Scale Component menu allows trapezoidal modification of wings or tails, and any components derived from them.

This is done by inputting the desired new area, aspect ratio, taper ratio, sweep, t/c, and/or dihedral. Even if you have extensively reshaped the wing such as modifying the planform, cropping the root, rounding the tip, or modifying the airfoil in some way, those changes will be preserved as the wing geometry is morphed to correspond to the new trapezoidal wing planform.

The powerful Shape Component menu features a number of ways to reshape a component's cross sections. They can be scaled, stretched, moved, and changed by direct point manipulation, and this can be done in side view, top view, and of course in cross section (rear) view. Section points can also be moved by selection of longitudinal control lines then direct manipulation in side or top view.

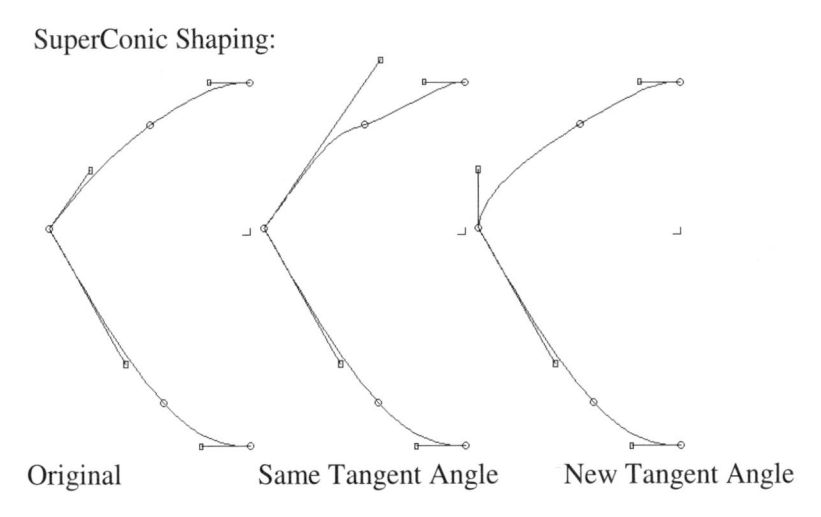

SuperConic Shaping:

Original   Same Tangent Angle   New Tangent Angle

Reshaping of a cross section can be done by directly moving the points, for either point-defined components or for SuperConic-defined components. For SuperConic sections, the control points are moved and the resulting lines are redrawn. When you move the tangent control points you are prompted if you really wanted to change the tangent angles or if you just wanted to move the point farther in or out along the existing tangent angle. Since these control points act as "magnets," moving the tangent control points farther out will preserve the tangent angle and reduce the curvature at that endpoint – the curve follows that tangent line more closely (see illustrations above). Alternatively, you can numerically enter the desired tangent angle. Finally, when you move a SuperConic endpoint, the adjacent tangent points are automatically moved as well, preserving the tangent angle.

Note that the screen is not automatically redrawn every time you move a point. Often designers want to "sneak up" on a geometry and would like to see incremental changes. Simply select Clear Background or Restore Background. Either one will clear the screen and redraw the section you are editing. Restore Background will also draw the rest of the airplane superimposed on the section you are working on.

Assemble Components

The locations and rotations of components can be changed by entering new values in Component Edit-Parameters, but a much more powerful method is found in the Assemble Component menu. This allows moving or rotating components by the selection of the desired change from a menu using Nudge, Move, or Jump as described above. The desired change can also be entered numerically, or you can go to Menuless Move/Rotate and "fly" the components into position using the same mouse/arrow/button controls described for Menuless Viewing below. In any case, this is all done in whatever view you are in when you select Assemble Components, even including perspective views (Why do that? It's fun!).

You can also select a number of different components to be moved or rotated all at once. You will be prompted for a "Leader" – whatever is done to the Leader is also done to the other selected components. This is especially useful to, say, move the landing gear which is probably composed of four or more separate components.

When you enter Assemble Component, Relative Viewing is disabled if it was active. Otherwise you would not see any motion of the Leader component – instead the rest of the airplane would appear to move or rotate in the opposite direction!

Undo Last Change

RDS-DLM has a powerful Undo capability. Every time a significant geometry change is made, the entire design is automatically saved to a temporary file (RDSDLM0.TMP to RDSDLM9.TMP). Undo recovers the most recent version, and can be repeated up to nine times if you have made so many changes to the design. Redo is also available.

During design work in RDS-DLM, you are actually working on temporary file RDSDLM0.TMP. When you Save in the Files and Analysis menu (below), you are merely renaming that temporary file to your project filename.

Note that these TMP files are full design layout module (.DSN) files, and in the event of a complete system crash or something similar you could copy any or all of them to a new file name, change the extension to DSN, and open them as usual in RDS-DLM. This should never be required because in such an event, RDS-DLM will detect the existence of un-deleted TMP files and offer to recover the most recent. Once you have done so, you can Undo to back up to earlier versions if desired.

Files & Analysis

A complete RDS-DLM file includes designed components plus optional "header" items such as Designer's name and Notes. The Files & Analysis menu allows storing, retrieving, and starting design files. Components can be copied from other files including a Component Bank where commonly-used items are stored for easy access.

Design Analysis creates a tab-delimited file (.TAB) of key design geometric information, and contains wing and tail data blocks, component information such as centroid location, and cross section perimeter & area. The TAB file can be read as text or as a Spreadsheet.

The RDS Aerodynamics and Weights Modules read the data from the TAB file to create or modify analysis inputs. To work properly, the components need to be identified as to type (wing, fuselage, wheel...). This is done with a code based on the SAWE group weight statement (RP8). In RDS-DLM you should use Comp Edit-Parameters to set the code for each component. Make sure there is one and only one Reference Wing.

Menuless Viewing

When menus are on the screen, they are used to select and change display views and options (see View Menu). These menus are easy to use and perfect for beginners and occasional users, but "experts" get tired of scrolling up and down through a series of menus to make changes in view type, rotation angle, or zoom. Menuless Viewing allows these to be done with a few simple keystrokes or mouse motions, but requires learning the "secret code" for each desired operation.

When Menuless Viewing is selected, the menu immediately disappears. Display control is then done with two hands, using mouse or arrow inputs plus the use of Ctrl and Alt keys. Mouse or arrow inputs move the aircraft image around on the screen. Holding down the Ctrl key causes mouse or arrow inputs to rotate the aircraft in pitch and roll. Holding down the Alt key while making left-right inputs yaws the aircraft. Alt-up/down zooms the display out or in.

To permit single-handed operation, yaw rotations can also be done with PageUp and PageDown buttons. Zoom is obtained using Insert and Delete buttons. Finally, the perspective viewing distance (normally set using menu) can be changed using Home and End buttons.

Menuless Operation Controls and Hot Buttons are listed on the next page – you may want to make a copy for reference if you intend to master this "experts" way of using RDS-DLM. Again, you don't have to learn these commands because all design and viewing capabilities can be done from the menus. But, Menuless Operation is faster and, as they say, way cool.

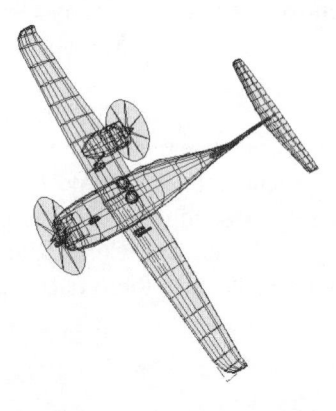

MENULESS OPERATION CONTROLS

```
Translation on screen:              Mouse or Arrows
Rotation (pitch & roll):     Ctrl-Mouse or Ctrl-Arrows
Rotation (Yaw):   Alt-Mouse, Alt-Arrow or PageUp/PgDown
Zoom out/in:      Alt-Mouse, Alt-Arrow or Insert/Delete
Perspective Distance farther/closer:           Home/End
Menu* (restore menus to screen):                  Alt-M
Exit from Design Module:                          Alt-X
 ScrollLock locks mouse to single X or Y axis input
 Ctrl-TAB toggles mouse off/on - Use before ALT-TAB
```

MENULESS OPERATION HOT BUTTONS

```
Orthographic (ends Perspective, keeps angles)         O
Isometric (ends Perspective, changes to Iso)          I
Views:         1=Side 2=Top 3=TopUp 4=Rear 5=RearSide
Three-Views (various display formats):     6 to 9 and 0

Zoom image to full screen                             Z
Numerical entry of rotation, perspective, & zoom      N
Mouse & key action - Slow/Default/Fast          S / D / F

Measure lengths and angles (iso & views 1-8 only)     M
Find Location (side, top, and rear views only)        L

Axis display (toggle: none/global/stations/local)     A
Trapezoidal Wing (toggle: none/wing only/all)         W
Mirror across global centerline (toggle)              %
Reflect across component centerline (toggle)          ^

Change Colors (toggle various combinations)           C
Show Rotation angles                                  R
Bitmap of screen image (always available)             B
Toggle between British (fps) & Metric (mks) Units     #
```

---

* Alt-M can be selected from within a menu to toggle to Menuless Operation. However, when you go back to the use of menus you may not find yourself exactly where you left off. Generally, when you Alt-M back to menu operation you return to the Main Menu or a top-level submenu.

DLM Main Menu Options

Several options already described in the View Menu are also available in the Main Menu Options for convenience, including Crop Screen and Toggle Colors. Other options include toggling between British Imperial units (fps) and Metric units (mks). You can also turn off the display of "what to do next" prompts that expert users will get tired of seeing again and again.

You can toggle between *Observer Mode* and *Pilot Mode*. These differ in the effect of mouse or arrow keys on position. In the default *Pilot Mode*, a right input moves the display or component to the right. In *Observer Mode*, a right input moves the display or component to the left – in other words, this is like moving a camera to the right.

One annoyance in many graphics packages is the difficulty computers have of keeping track of the aspect ratio of the display screens and printers. Often, squares are not truly square as seen on the screen or in a printout, despite the promises of the hardware and operating system vendors.

RDS-DLM solves this problem in a brute-force fashion. It stores two vertical-axis "fudge-factors," one to provide a square screen display and one to provide a square printed output. These may be the same, but usually are different. You select the active one with the Image XY Ratio option. Always select the Printer XY Ratio factor before making bitmap images so that they have the correct proportions.

Either value can be changed. RDS provides a square image for test and instructs you how to measure it and input the appropriate factor. This needs to be done only once, and only if you really want to get the images exactly correct.

FUNCTIONAL ANALYSIS MODULES

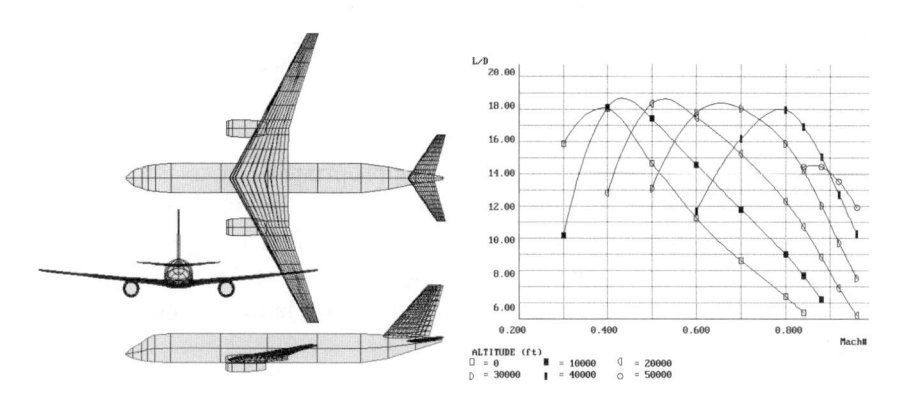

FUNCTIONAL ANALYSIS OVERVIEW

Air vehicle functional analysis includes aerodynamics, stability & control, weights, and propulsion. In RDS these are largely based upon the classical analysis methods described in Dr. Raymer's widely-used textbook, *Aircraft Design: A Conceptual Approach* (4th edition). These methods have been shown to provide reasonable and reliable results for normal aircraft and, when properly applied by a knowledgeable engineer, can give good results for a wide variety of unusual designs. RDS functional analysis has been correlated with a wide variety of actual aircraft including the Airbus A-321, the Lockheed L-1011, Global Hawk, the Cessna Caravan, the T-38/F-5, Dark Star, early JSF designs, and many more, and has given surprisingly good results considering the relatively few inputs and small amount of time required for analysis. Yes, better methods are available, but don't be surprised if those methods require ten times the effort for a few percent better answer!

Inputs for vehicle analysis are defined by the user in spreadsheet-like matrices and saved in data files with the appropriate filename extension (.DAA for aero, .DWT for weights, and .DPR for propulsion). Required inputs are indicated for each row and column of the data matrices using the variable names employed in Dr. Raymer's book. Equation numbers and figures provided below are also from the textbook.

One peculiarity about the analysis input screens – the column names down the left side of the screen apply only to the "active" column, i.e., the column with the highlighted input item. If the active column is not the first column it is easy to get confused and think that a value in the first column is a really strange input. You will soon get used to it!

Once an input file is created and the appropriate data is entered, *Do Analysis* can be selected from the menu buttons at the bottom of the screen. It is wise to review the analysis results first by selecting *Show Summary Results*, then run *Analysis* again and select *Load Results into Aircraft Data File*. Then, go to the Aircraft Data File to see and graph the full results.

Note that RDS does NOT automatically save the analysis results to a permanent file. This is only done from the Aircraft Data File Module where the data from Aerodynamics, Weights, and Propulsion are collected for review and application. You must go to that module and select *Save* to do so. This RDS feature is deliberate – you should carefully review the results before saving them or using them in Sizing, Performance Analysis, or Optimization and Trade Studies.

AERODYNAMICS MODULE

The RDS Aerodynamics Module estimates subsonic and supersonic parasite drag, drag due to lift, lift curve slope, and maximum lift from a user-defined input matrix. Subsonic parasite drag is estimated by the component buildup method (eqs. 12.24-12.39 in Dr. Raymer's textbook). Supersonic wave drag is determined by the equivalent Sears-Haack technique (eq. 12.46). Transonic drag is determined by empirical fairing between Mdd (figs. 12.26 and 12.27) and the supersonic wave drag, using a cubic polynomial curve-fit (fig. 12.29).

Drag due to lift is calculated by the leading-edge suction method using $C_{L\text{-alpha}}$ calculated with equation 12.6 and fig. 12.6. The input design lift coefficient ($C_{L\text{-design}}$) is used to select the leading edge suction schedule from Figure 12.35.

For subsonic, thick-wing designs, Figure 12.35 is automatically modified during RDS analysis to provide 93% leading edge suction at all lift coefficients below design CL. This is common industry practice.

For high aspect ratio wings, the leading edge suction schedule actually becomes a function of aspect ratio, with percent suction increasing for very high aspect ratio wings. RDS handles this by automatically adjusting the default suction schedule in fig. 12.35 so that an equivalent Oswald's Span Efficiency Factor ("e") at the design lift coefficient is no less than the value stored in file RDSCONST (e=.8 unless you change it). This correlates well with actual aircraft data. No additional user inputs are required, just the wing aspect ratio.

For an advanced fighter with automatic maneuvering flaps, their contribution should be included in the Aircraft Data File matrix both as an increase in maximum lift and a reduction in drag-due-to-lift factor "K." RDS estimates the reduction in K for such an aircraft by shifting the leading edge suction schedule by a C_L of 0.2, which is activated by entering a *negative* value for design lift coefficient in the Aerodynamics input matrix. Enter the lower design lift coefficient value preceded by a minus sign. For example, if you enter (-.4) as the design lift coefficient, RDS will use the (.4) design C_L suction schedule up to (.4), maintain the peak value to (.6), then follow the (.6) design C_L schedule for higher values. This is a fairly rough approximation, but gives reasonable results for early design studies.

Maximum lift is calculated using Figures 12.8 and 12.9, also from DATCOM charts. Those charts and their RDS implementation are considered reasonable only for moderate to high aspect ratio configurations. For very low aspect ratio configurations (<<3) this method underestimates maximum lift because it ignores the additional vortex lift contribution that should be expected. For a better answer you can estimate maximum lift yourself then enter it in the appropriate matrix in the Aircraft Data File (.DAT). Or, you can use the RDS estimate knowing that it will be somewhat conservative.

If you choose to enter the maximum lift directly in the Aircraft Data File, make sure you leave the airfoil maximum lift coefficient in the

Aerodynamics Module input file as zero (0). This stops RDS from doing its own maximum lift calculation and overwriting your input value.

Maximum lift as calculated and stored in the Aircraft Data File $C_{L\text{-max}}$ array is for the flaps-up condition and is used for up-and-away performance calculations. The flaps-down maximum lift for takeoff and landing calculations can be treated in one of three ways.

By selecting and completing an input column for leading and/or trailing edge devices in the Aerodynamics Module, RDS will calculate reasonable takeoff and landing drag polars including high-lift devices. These polars are stored in the Aircraft Data File as Alternate Drag Polars #1 and #2, along with Polar #3 which is a low-speed flaps-up polar. If you also select a landing gear input column, the gear drag will be added to the polars.

Alternatively, the user can directly enter Alternate Drag Polars for takeoff and landing into the Aircraft Data File. This is often done when data exists for a similar configuration – it is normal in industry to trust adjusted real data more than bottoms-up analytical estimates in this hard-to-predict parameter.

In either case, the appropriate Alternative Drag Polar number must be indicated in the Performance Analysis inputs (see below). This causes the performance analysis to ignore all other aerodynamic data in favor of the alternative polar selected.

A third option for entering maximum lift ($C_{L\text{-max}}$) is to enter it directly in the Performance Analysis inputs in the selected columns. This overrides the other options.

Longitudinal stability and control are evaluated using first-order methods from the DATCOM and other sources. Stability and control calculation is optional, and is obtained by selecting a Stability and Control data column in the Aerodynamics Module, entering the required data, and then running analysis. RDS calculates the neutral point, pitching moment derivative, and other stability parameters. Then it displays a trim plot for the input speed and altitude, followed by a graph of neutral point, static margin, and $C_{M\text{-}\alpha}$ as functions of Mach number. For subsonic aircraft (M<.7), both stick-fixed and stick-free results are presented.

If you are storing aerodynamic results in the Aircraft Data File and you have a Stability and Control data column, trim drag as a function of Mach number and lift coefficient can be calculated and stored as an adjustment to the induced drag factor (K). Trim drag estimation includes the change in lift on the wing, the induced drag of the tail at the trim elevator deflection, the rotation of the tail lift vector due to wing downwash, and the parasitic drag impact of the elevator deflection.

This total increase in drag is used to adjust the drag-due-to-lift factor "K" such that the total drag-due-to-lift including the effect of trim can be subsequently calculated just by multiplying this adjusted K by C_L^2. This speeds up execution during iterative sizing and optimization runs. However, when you then graph the resulting K matrix it may look strange compared to the untrimmed K matrix – this is to be expected.

When creating a new aerodynamic input file, the user first selects from a menu the desired types of aerodynamic components such as fuselage, tails, and canopy. Note that an input column for a reference wing component is automatically provided, and its area will be the reference area for all aerodynamic coefficients. You cannot analyze a design without a wing, so when analyzing a blimp or rocket with no wing, you should enter a "fake" wing of the desired reference area, but with wetted area set to zero so that the parasitic drag of the fake wing isn't included in the analysis.

You may not select additional wings. Instead, analyze additional wings as "horizontal tails." These additional wings will only be used to calculate parasite drag and will have no affect on lift or drag due to lift.

For components and drag contributions not available in the list, you may select the arbitrary input of D/q as a function of Mach number. This is typically done for an external store or the drag penalty for an unusual radome or refueling probe.

Required geometric information such as wetted area can be automatically determined from the RDS design layout module or can be input by the user. (See the chapter on the design layout module for procedures for geometric analysis and data transfer).

```
DR3.DAA          AERO DATA      WING          HORZ TAIL      FUSELAGE       CNPY/FAIR
Max V or M#         2.0000       1.0000          1.0000        1.0000          1.0000
Max Altitude    50000.0000     294.0000         92.0000      588.0000         39.0000
% Laminar           0.0000     215.0000         92.0000       45.2000         13.9000
k/10^5 ft           3.3300       3.5000          4.0000        5.5000          2.0000
%Leak&Protub        6.0000       3.5000          4.0000        1.0000          1.0000
Amax-aircrft       17.0700       0.2500          0.3400        0.0000          0.0000
length-eff         45.2000      38.0000         30.0000        0.0000          0.0000
Ewd                 2.0000       0.0600          0.0600        0.0000          0.0000
CL-cruise           0.2100       1.2800          1.2800        0.0000          0.0000
%WD Flatness        0.0000       1.0000          1.0000        0.0000          0.0000
1=Alt.Suction       0.0000       0.4000         28.4000        0.0000          0.0000
MnvrFlapdClmax      0.0000       1.6400          0.0000        0.0000          0.0000
set F*(Sexp/S)      0.0000       1.0000          1.0000        1.0000          1.0000
```

Figure Two. Aerodynamic Inputs

Figure Two shows the aerodynamic input screen with different rows active for input. The AERO DATA column is active, as indicated by the highlighted data item (**50000**). The row headings to the left indicate the proper inputs for the active column, and change as you move the cursor across the screen to different columns.

The first item in most aerodynamic data columns is the number of components of the type described in the column. RDS already counts the left and right wing (or tail) as one component, so unless you are designing a biplane, a 1 should be entered. If you are designing a biplane you could enter a 2, and the parasite drag would be properly estimated. However, RDS does not contain aerodynamic analysis for K and maximum lift of a biplane, so you must estimate those yourself then type the results into the Aircraft Data File.

The term *%WD Flatness* permits adjusting the estimated magnitude of the typical reduction in Wave Drag coefficient past Mach 1.2, as calculated by the statistical middle terms in equation 12.46 of the textbook. This statistical term is based upon earlier supersonic jets which typically had rather large wings compared to modern aircraft, so this equation may indicate an excessive reduction in wave drag coefficient for a more modern design. Entering 0 for *%WD Flatness* uses equation 12.46 as written. Entering 100 causes the resulting wave drag curve to be completely flat (100% flat), with no reduction past Mach 1.2. This occurs sometimes when the wing is rather small relative to the overall cross section area.

A reasonable value for this term can be determined by comparing RDS results with data for an existing aircraft similar to your concept. For a modern fighter, 50% seems to be reasonable. Some companies and government organizations always apply a conservative 100% flatness factor (i.e., no adjustment) even when more sophisticated drag codes indicate otherwise!

A key parameter in the analysis of wing lift curve slope as well as induced drag is the product {F*Sexp/S}, which is the fuselage lift factor times the ratio of exposed to reference wing area (see Dr. Raymer's book, equation 12.6). If this product is greater than one, it implies that the total lift of the actual wing plus the fuselage area covering the wing is greater than the lift of the trapezoidal wing itself. This can happen, but rarely. For example, a wing with a large strake may generate more lift than its trapezoidal planform would indicate. However, it is unlikely and RDS warns you should it occur.

RDS allows you to override this lift contribution estimate {F*Sexp/S} by inputting your own value in the first column input item *Set F*Sexp/S*. Setting this input to something other than zero forces that value, otherwise the calculated product is used. As a typical, conservative limit, 0.95 can be used. The maximum lift calculation is also corrected by this input term.

A critical input for the analysis of parasitic drag is the assumed percentage of laminar flow. This is used to determine the flat-plate skin friction coefficient, interpolating between turbulent and laminar values. For most supersonic aircraft, a reasonable assumption as to laminar flow at supersonic speeds is simple - there isn't any! Therefore, RDS normally uses turbulent skin friction coefficients at supersonic speeds regardless of the input percent laminar flow. However, recent research offers the hope of attaining some laminar flow at supersonic speeds. By inputting the percent laminar flow as a *negative* number in the aerodynamics inputs, the input amount of laminar flow will be assumed at both subsonic and supersonic speeds (good luck in actually attaining this!).

One confusing aspect of the aerodynamic inputs must be noted. In the Fuselage drag column, effective diameter "d" for the fuselage is required. This is the full effective diameter, and is used to determine the form factor for subsonic parasitic drag. In the first column, though, Amax is input.

Amax is the total maximum cross section area of the WHOLE AIRCRAFT, including the wing, but does NOT include the cross section area of the capture area of a jet engine. This term is used to estimate supersonic wave drag. The air that goes "down the hole" (i.e., into the engine) doesn't affect the wave drag.

Fudge factors permit arbitrarily reducing or increasing drag by some value input by the user. The use of fudge factors can be toggled on or off from the menu area, so that you can make an aerodynamic calculation ignoring the fudge factors, without having to reset them all equal to 1.0.

When executing the aerodynamics analysis, RDS provides the option of loading the data into the Aircraft Data File, and if selected will load the calculated parasite drag, drag due to lift factors (K), and maximum lift coefficients in the working version of the Aircraft Data File. Upon exiting the Aerodynamics Module and entering the Aircraft Data File Edit Module, the analysis results may be printed in tabular format or viewed using the Graph option.

RDS does NOT automatically save these results to a disk file, and you must go to the Aircraft Data File Edit Module and select Save to do so. Also, if you wished the aerodynamic results to be appended to a complete aircraft data file you are building, that file must be read into the working file area BEFORE you run the aerodynamic analysis. Do this by entering the Aircraft Data File Edit Module and giving the appropriate filename before running the aerodynamic analysis.

Figures Three and Four show typical results. If desired, you can select a full printout of the aerodynamic calculations at all velocities and altitudes, but this consumes a lot of paper. Alternatively, you can print the full results to a file.

These aerodynamic results stored in the Aircraft Data File, along with the required propulsion and weights data as described below, may then be used for sizing and performance calculations.

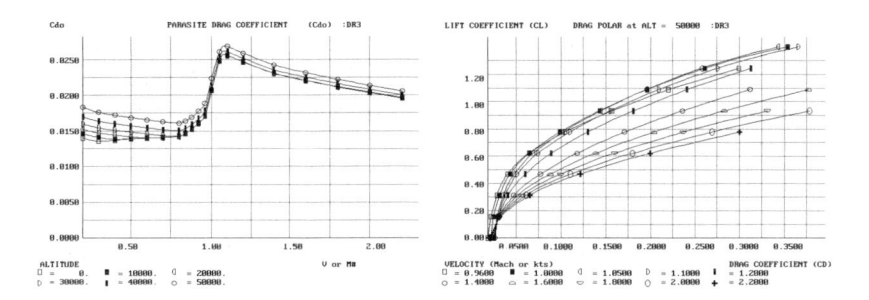

Figure Three. Aerodynamic Results   Figure Four. Aero Results, Cont

WEIGHTS MODULE

Weights and balance are estimated statistically from inputs defined in a spreadsheet-like matrix, after selection of aircraft type (fighter, transport/bomber, or general aviation). Equations and variable name definitions are provided in Dr. Raymer's textbook.

You *must* give all relevant inputs if a default value is not provided in the array. While RDS checks for divide-by-zero, you will get nonsense results if a parameter is left out. You may leave all zeros if your aircraft doesn't even have a particular component (such as a horizontal tail), but for other components you must set all parameters including constants which are 1.0 for your design (for example, you must set Kdw=1.0 for the wing of a non-delta-wing design). READ THE VARIABLE DESCRIPTIONS CAREFULLY as you input them into RDS!

Note that the statistical weights are estimated from an input design gross weight (Wdg). This may or may not be the same as the design takeoff gross weight (Wo), which is the total weight used for the group weight statement to calculate fuel available. Inputs are required in RDS for both Wo and Wdg, for fighters and transport aircraft. For fighters especially, it is common to define Wdg as the aircraft weight after burning off some of the design fuel. A typical Wdg is defined as the aircraft weight carrying some specified payload, with 50-60% of maximum internal fuel. This may be only about 85% of Wo.

In the last column of weights entries, a fudge factor can be applied to the total previously calculated empty weight. This provides an Empty Weight

Margin which reduces the project risk. Typical values are between 5% (input as 1.05) and 10 % (1.1). Upon weights analysis, the empty weight margin will appear as an addition to miscellaneous empty weight in the weight statement.

Fudge factors permit estimation of the weight impact of non-standard materials and other emerging technologies. Fudge factors can also be used to calibrate the RDS weights equations to some known design. The use of fudge factors can be toggled on or off from the menu area, so that you can make a weight calculation ignoring the fudge factors, without having to reset them all equal to 1.0.

For general aviation aircraft, a simplified but often-used method for weights estimation can be employed in RDS instead of the statistical equations described in the textbook. Component weight ratios, such as pounds per square foot (kg/sq m) can be used to estimate wing, tail, and fuselage weights. These inputs, if used, completely override all other weights inputs (except that fudge factors are still applied).

Center of gravity is determined from input component locations, and is summed for empty weight, zero-fuel weight, and takeoff gross weight. Typical weights inputs and summary weights result are shown as Figures five and six.

RDS uses an exponential coefficient to estimate the variation of empty weight fraction with takeoff weight during sizing. This Weight Coefficient C corresponds to the slope of the log of empty weight fraction plotted versus the log of takeoff weight, as defined in table 3.1 and Figure 3.1 in Dr. Raymer's textbook. You can use a C value from table 3.1 or develop your own regression using data on aircraft similar to the one you are designing. This value must be a small negative number such as (-.05), and is stored in the first column of the first matrix of the Aircraft Data File (*Wt Coeff*).

If selected during *Do Analysis*, RDS will load the weights results into the working version of the Aircraft Data File. This includes the calculated Empty Weight as well as the input Takeoff Gross Weight, Crew Weight, Passenger Weight, Cargo (payload) Weight, Misc. Useful Load, and empty weight fraction sizing coefficient. Fuel weight is not stored because the fuel weight is always calculated by RDS as the remainder when empty weight and useful load are subtracted from takeoff gross weight. Fuel weight available is shown in the Aircraft Data File but cannot be changed except by changing one of the other weight values. No cheating!

Remember, RDS does NOT automatically save these results to a permanent file until you select *SAVE* in the Aircraft Data File Edit Module.

These weights results stored in the Aircraft Data File, along with the required aerodynamic and propulsion data, may then be used for sizing and performance calculations.

```
DR3.DWT            AC DATA        WING       HOR. TAIL    VERT TAIL    FUSELAGE
Wo   (TOGW)      16480.0000      23.3000      39.2000       0.0000      21.7000
Wdg  (flight)    16480.0000       1.0000       4.7000       0.0000       1.0000
Nz   (ultimate)     11.0000       1.0000       6.5000       0.0000      39.0000
Sw                 294.0000       3.5000      90.0000       0.0000       4.0000
M design max         1.8000       0.0600       0.0000       0.0000       5.4000
#Engines             1.0000       0.2500       0.0000       0.0000       0.0000
1=fixsize eng        0.0000      30.0000       0.0000       0.0000       0.0000
1=fix We-misc        0.0000      72.0000       0.0000       0.0000       0.0000
```

Figure Five. Weights Inputs

```
GROUP WEIGHT STATEMENT  - FIGHTER/ATTACK
STRUCTURES GROUP         4526.2       EQUIPMENT GROUP              3066.7
  Wing              :    1459.4         Flight Controls    :        655.7
  Horiz. Tail       :     280.4         Instruments        :        122.8
  Vert. Tail        :       0.0         Hydraulics         :        171.7
  Fuselage          :    1574.0         Electrical         :        713.2
  Main Lndg Gear    :     631.5         Avionics           :        989.8
  Nose Lndg Gear    :     171.1         Furnishings        :        217.6
  Engine Mounts     :      39.1         Air Conditioning   :        190.7
  Firewall          :      58.8         Handling Gear      :          5.3
  Engine Section    :      21.0         MISC EMPTY WEIGHT  :       1000.0
  Air Induction     :     291.1       TOTAL WEIGHT EMPTY   :      10947.2

Propulsion GROUP         2354.3       USEFUL LOAD GROUP            5532.8
  Engine(s)         :    1517.0         Crew               :        220.0
  Tailpipe          :       0.0         Fuel               :       4422.8
  Engine Cooling    :     172.0         Oil                :         50.0
  Oil Cooling       :      37.8         Payload            :        840.0
  Engine Controls   :      20.0         Passengers         :          0.0
  Starter           :      39.5         Misc Useful Load   :          0.0
  Fuel System       :     568.0       TAKEOFF GROSS WEIGHT        16480.0

EMPTY CG=  23.8    LOADED-NO FUEL CG=  23.4   GROSS WT CG=  23.1

EMPTY WEIGHT SIZING COEFFICIENT (for small changes):  C=-.193
```

Figure Six. Weights Results

PROPULSION MODULE

Propulsion analysis includes installation analysis for jet engine thrust and specific fuel consumption, and propeller thrust and specific fuel consumption for piston-prop engines. Turboprop engine analysis includes both propeller analysis and inclusion of the additional thrust due to residual jet exhaust.

Jet engine installation analysis takes an already-saved file of manufacturer's uninstalled engine data and applies corrections per chapter 13 of Dr. Raymer's textbook. The uninstalled engine data (thrust and specific fuel consumption) is entered using the Aircraft Data File Edit Module. When this file is saved, it must be given a filename ending in .ENG or it will not be listed as available when you need it.

After this uninstalled engine file is created and saved, the Propulsion Module is selected and a propulsion input file is created. An RDS spreadsheet-like matrix is used to input required parameters such as reference and actual inlet pressure recovery (vs. Mach), bleed coefficient, and inlet drag (vs. Mach). Defaults are provided for many values, such as the MIL-E- 5008B reference pressure recovery schedule (eq. 13.5), and typical actual pressure recovery schedules for various types of inlets. These defaults may be overridden simply by entering new values in the matrix.

For propulsion analysis of a non-rubber, or "fixed-size" engine, enter a zero (0) in the "Thrust-net" entry item. RDS will use the given engine without scaling, and will determine the resulting installed thrust. This is used for design with an actual engine.

Typical jet propulsion analysis inputs and results are shown below.

DR3.DPR	Misc Data	Mach -----	P1/PoREF	Mach -----	P1/PoACT
Thrust-net	16150.4000	0.2000	1.0000	0.2000	0.9700
SFC Fudge	0.8000	0.4000	1.0000	0.4000	0.9700
Acapture	3.8300	0.6000	1.0000	0.6000	0.9700
C-bleed	0.0000	0.8000	1.0000	0.8000	0.9700
bleed ratio	0.0000	1.0000	0.9700	1.0000	0.9700
Nozzle Cd	0.0150	1.2000	0.9617	1.2000	0.9680
Amax-nacelle	16.9000	1.4000	0.9489	1.4000	0.9600
(n/a)	0.0000	1.6000	0.9335	1.6000	0.9450
(n/a)	0.0000	1.8000	0.9162	1.8000	0.9120
(n/a)	0.0000	2.0000	0.8972	2.0000	0.8300
(n/a)	0.0000	2.2000	0.8769	2.2000	0.7200
(n/a)	0.0000	2.4000	0.8554	2.4000	0.6000
(n/a)	0.0000	0.0000	0.9700	0.0000	1.0000

Figure Seven. Jet Propulsion Inputs

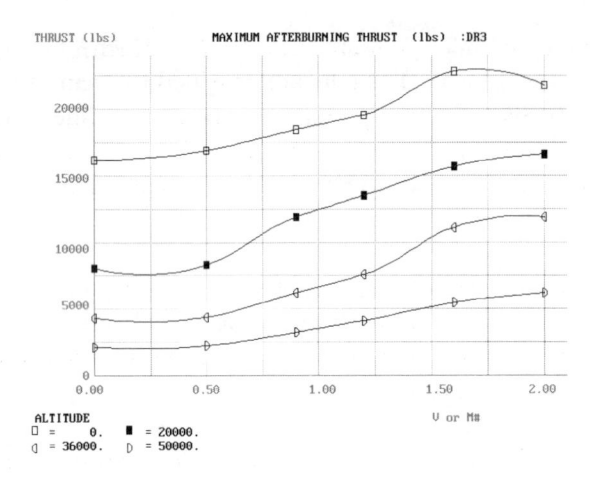

Figure Eight. Jet Propulsion Results

Propeller propulsion analysis does not require (or use) an uninstalled engine file, and instead calculates thrust and specific fuel consumption strictly from inputs in the Propulsion Module input file. These inputs include horsepower and brake specific fuel consumption as functions of altitude, propeller efficiency as a function of Advance Ratio (J) and Power Coefficient (Cp), and static thrust coefficient ratio as a function of power coefficient. Default values are provided for various types of propellers, or you can enter new values from some other source.

Engine power is input as a function of altitude at maximum continuous cruise power. If only one value is given, RDS assumes a normally-aspirated (non-supercharged) engine and degrades power with altitude following equation 13.9 from Dr. Raymer's textbook. The single value given for maximum (takeoff) power is ratioed in parallel with the given cruise power, although generally maximum power is not used at high altitudes. Power specific fuel consumption for takeoff power setting is input as a ratio to the cruise setting.

If a piston-prop engine has power that is substantially a function of forward speed for some reason, it should be treated as a turboprop with zero residual thrust (see below).

Propeller thrust calculation interpolates the propeller efficiency tables to calculate gross thrust for the input engine power. This includes a blockage adjustment that acts to reduce J for the effect of the nacelle behind the propeller. A tip Mach number correction is employed, then scrubbing, cooling, and misc. drag adjustments are added. A static thrust value is estimated from the table of Ct/Cp, but is adjusted as required to ensure that the static thrust doesn't exceed the 25 kt thrust by more than 10%, and doesn't fall below the 50 kt value. Empirical fairing is then used between the static value and the forward flight thrust results.

Note that the propeller analysis creates exactly the same type of data as the jet installation analysis, namely thrust and specific fuel consumption as a function of velocity and altitude. This permits the sizing and performance modules to merely read the values when required rather than recalculate them every time they are needed. Typical propeller inputs and results are shown in Figures Nine and Ten.

RDS handles piston-propeller engine data under takeoff conditions (maximum horsepower, full rich mixture) by storing it in the first two engine matrices of the Aircraft Data File). The third and fourth matrices store thrust and specific fuel consumption under leaned-out cruise conditions. One special input required for this in the Propulsion Module is the ratio of fuel consumption from max-rich to lean mixture setting ($c_{max}/c_{economy}$), which is typically about 1.15 for regular engines and about 1.55 for turbocharged engines.

Note that in the mission and performance modules the user should indicate maximum power, full-rich operation by a throttle setting of 999 (like afterburner for a jet), and indicate a maximum-continuous cruise throttle setting as a percentage somewhat less than full power, perhaps 80%.

For turboprop engines, a previously-saved file of manufacturer's uninstalled engine data provides (horse)power and power-specific specific fuel consumption which are used for calculation of propeller thrust and thrust-specific specific fuel consumption as described above. The effects of residual thrust are then added to the propeller thrust. The engine data must be entered using the Aircraft Data File Edit Module. When this file is saved, it MUST be given a filename ending in .TBP !

These propulsion results stored in the Aircraft Data File may then be used, along with the aerodynamics and weights data, for sizing and performance calculations.

```
GA-DPR.DPR     Misc Data     Altitude ----- Cont.Power     Altitude ----- C Econ
MaxPower @ SL   425.0000        0.0000       425.0000        0.0000       0.5000
MaxP SFCfudge     1.0000     20000.0000      425.0000     30000.0000      0.5000
rpm-MaxCruise  3200.0000     30000.0000      200.0000        0.0000       0.0000
PropDiameter      6.8000        0.0000         0.0000        0.0000       0.0000
Cfe (counts)     20.0000        0.0000         0.0000        0.0000       0.0000
Swet-washed     500.0000        0.0000         0.0000        0.0000       0.0000
Fudge-misc        2.0000        0.0000         0.0000        0.0000       0.0000
V-max           350.0000        0.0000         0.0000        0.0000       0.0000
Max Alt       30000.0000        0.0000         0.0000        0.0000       0.0000
C-max/C-Econ      1.2000        0.0000         0.0000        0.0000       0.0000
Blade t/c         0.0000        0.0000         0.0000        0.0000       0.0000
Nacel x-area      0.0000        0.0000         0.0000        0.0000       0.0000
    (n/a)         0.0000        0.0000         1.0000        0.0000       1.0000
```

Figure Nine. Propeller Inputs

```
STATIC THRUST Analysis AT SEA LEVEL        (max continuous power setting)
   POWER COEFFICIENT :    Cp =    0.0445              POWER =    425.00 Hp
   THRUST-POWER RATIO: Ct/Cp =    2.7804       STATIC THRUST =   1731.66 lbs-f

FORWARD FLIGHT THRUST Analysis AT SEA LEVEL
 V (kts)    J       Eta-P    Gross Thrust   Cooling Drag   Misc Drag
   50.    1.0231    0.5207    1441.2692       11.2745       1.4428
  100.    1.0314    0.7001     968.8712       22.5489       5.7711
  150.    1.0452    0.7973     735.6065       33.8234      12.9849
  200.    1.0642    0.8255     571.2233       45.0979      23.0842
  250.    1.0881    0.8159     451.6627       56.3724      36.0691
  300.    1.1167    0.7793     359.4847       67.6469      51.9395
  350.    1.1495    0.7202     284.7906       78.9213      70.6954
```

Figure Ten. Propeller Results

AIRCRAFT DATA FILE MODULE

As described in the manual overview, the Aircraft Data File Module is the analytical heart of RDS. It allows you to review and revise the technical data that defines the aircraft for subsequent sizing, range, and performance calculations. The Aircraft Data File includes calculated C_{D0}, K, $C_{L\text{-max}}$, installed engine thrust and specific fuel consumption, and other sizing inputs such as the As-Drawn Takeoff Weight and the Payload Weight. Parameters are stored in matrix format as a function of one or two variables (such as velocity and altitude).

Data can be directly entered if it is available from some other source, or else the RDS Aerodynamics, Weights, and Propulsion Modules can be used to create the data and load it into the Aircraft Data File.

When using these analysis modules to build the Aircraft Data File remember that the *Load Results...* option copies the analysis results to the working file. If you wish to add new analysis results to a project aircraft data file you are creating, that file must already be in the working file. If you have just started RDS you must therefore first get the project Aircraft Data File from the disk by selecting the Aircraft Data File Module and entering the file name. After calculating data in an analysis module and loading it to the Aircraft Data File, you must *save* that Aircraft Data File before exiting RDS or the data will be lost!

The Aircraft Data File consists of seven arrays as follows. The first array is really a collection of columns, as described below. The rest of the arrays are matrix tables of one or two variables such as speed and altitude.

```
      -Weights, T/W, W/S, Stores Drag, & Alternate Drag Polars
               (individual entries in a multi-column format)

      -Cd0 (PARASITE DRAG COEFFICIENT)
               (vs. altitude and velocity or Mach#)

      -K (INDUCED DRAG FACTOR)
               (vs. Cl and velocity or Mach#)

      -CL MAX  (MAXIMUM LIFT COEFFICIENT)
               (vs. velocity or Mach#)

      -MAXIMUM AFTERBURNING THRUST (or max power propeller thrust)
               (vs. altitude and velocity or Mach#)

       -C AT MAXIMUM AFTERBURNING THRUST (or max power propeller SFC)
               (vs. altitude and velocity or Mach#)

      -MAXIMUM DRY THRUST    (or propeller thrust)
               (vs. altitude and velocity or Mach#)

      -C AT MAX DRY THRUST   (or propeller thrust SFC)
               (vs. altitude and velocity or Mach#)
```

Figure Eleven shows the input screen for values such as the takeoff T/W and W/S, the as-drawn takeoff and empty weights, the crew, cargo, passenger, and oil weights, and the value for weight of the miscellaneous useful load.

The statistical exponential coefficient to be used for variation of empty weight fraction with takeoff weight during sizing is also entered in the first column of this array. This *Wt Coeff* is described in the Weights Analysis section above, and must be a negative value. Number of engines, aircraft maximum load factor, and dynamic pressure are entered in the second column of this matrix.

As an alternative to the input of T/W and W/S, thrust per engine (T) and wing area (S) can be entered in the same data positions. Thrust is given in pounds {kN}, as maximum sea level static thrust per engine. This need not be the same value given in the thrust matrix - RDS will automatically scale all thrust values to match the thrust provided in this data item, and then multiply times the number of engines. If thrust is given instead of T/W, then wing area (S) MUST also be given (in square feet or meters). You CANNOT mix the inputs, such as T and W/S. Also, giving actual thrust per engine instead of T/W forces use of a fixed-size-engine during sizing (see below).

```
              Wts Data    Misc Data      Mach ----- StoreD/q     # 1   :Cl
T/W (TO)       0.98000     1.00000     0.00000     0.00000      0.00000
W/S (TO)      56.05440     7.33333     0.00000     0.00000      0.00000
Wo-Drawn   16480.00000  2200.00000     0.00000     0.00000      0.00000
We-Drawn   10947.22000     0.00000     0.00000     0.00000      0.00000
Wcrew        220.00000     0.00000     0.00000     0.00000      0.00000
Wcargo       840.00000     0.00000     0.00000     0.00000      0.00000
Wpassngr       0.00000     0.00000     0.00000     0.00000      0.00000
Wmisc UL       0.00000     0.00000     0.00000     0.00000      0.00000
Woil          50.00000     0.00000     0.00000     0.00000      0.00000
Wt Coeff C    -0.19343     0.00000     0.00000     0.00000      0.00000

     Available Wfuel=    4422.78 lbs-m
```

Figure Eleven. T/W, W/S, Weights and Other Data

If either takeoff T/W or T is given in the first column, the engine thrust values in the thrust matrix will be scaled assuming that the first (upper-left) entry in the thrust matrix is the sea level static thrust (afterburning if available, otherwise dry power). Thus, you must have zero altitude and zero velocity rows and columns in your data matrix. This is true for jet or prop aircraft.

To perform sizing and performance calculations using a fixed-size, existing engine, you may select fixed-size engine sizing from the Options menu in the Sizing & Mission Analysis Module. Mission sizing using a fixed-size engine is also forced if you give the actual thrust per engine (sea level static) in this first column. Then the thrust data stored in the Aircraft Data File matrices is scaled to match the sea level static value provided, times the number of engines. As mentioned before, when providing T instead of T/W, you must also provide wing area S instead of W/S.

You can also force mission sizing with a fixed-size engine by providing a "0" value for T/W in the first matrix above. This causes RDS to use the given engine data exactly as input in the Aircraft Data File thrust arrays. The total thrust is then the number of engines times the thrust value stored in the thrust matrix (for a given speed and altitude). When T/W is set to zero, you cannot input wing area S directly, and must instead input W/S. This is obvious from the input labels when you set T/W to zero.

The use of T/W=0 to indicate a fixed-size engine analysis means that you cannot analyze a glider or sailplane using RDS by inputting T/W=0. To analyze a glider or sailplane it is necessary to trick RDS, giving an extremely small value for thrust in the thrust matrix (.000001, for example), and some small but nonzero T/W value (such as .1). Don't give an extremely small value for T/W because, for reasons of numerical precision, RDS tests for a very small T/W rather than T/W exactly equal to 0.0 when looking for T/W=0.

This input screen also includes two columns defining stores drag (D/q per store) as a function of Mach number. Stores drag refers to droppable stores, such as bombs, that you did *not* include in the C_{D0} matrix as described below. The Sizing and Performance Modules have inputs for the number of droppable stores (which can change during the mission). During analysis, the number given is multiplied by the D/q values in the matrix. If your aircraft is carrying and dropping different types of stores, the value stored should be the average D/q.

For special situations such as takeoff and landing where the strong effect of flaps increases both lift and drag, you can directly input and use aircraft drag polars (typically taken from actual data for a similar aircraft). These are input and stored in the Weights, T/W, W/S, etc... matrix as Alternative Drag Polars number 1 to 7. The Sizing and Performance Modules allow you to select use of an input Alternative Drag Polar by its number. The Aerodynamics Module can also estimate such takeoff and landing alternative drag polars and store them in the Aircraft Data File.

```
    Mach # =        0            .5           .9          1.2          1.6
 alt=  0 ft   16150.40000  18761.45000  20579.26000  21396.93000  25003.91000
 alt= 20000    8075.20000   9234.61000  13279.51000  14787.37000  17205.95000
 alt= 36000    4306.77000   4887.02000   6948.56000   8295.25000  12218.50000
 alt= 50000    2153.39000   2503.15000   3626.47000   4515.98000   6011.89000
```

Figure Twelve. Typical Data Matrix

Figure Twelve illustrates a matrix data array typical of the rest of the Aircraft Data File, in this case for thrust. Others include parasitic drag, induced drag, $C_{L\text{-max}}$, and SFC. On the screen, the current cursor location is visible as a brighter number. To enter a new value, just move the cursor to the old number with mouse or arrows and type in the new value.

You can create these matrices simply by typing in the data yourself, or by using the Aerodynamics, Weights, and Propulsion analysis modules. These can store their results directly in the Aircraft Data File.

If the Aerodynamics, Weights, And Propulsion Modules were all used (properly!) for analysis and the results were loaded into the Aircraft Data File, then Sizing and Performance Analysis can be run. Saving the Aircraft Data File first is a good idea – sometimes a bad input will require restarting RDS and the data will be lost.

Aircraft drag is normally calculated from the C_{D0} and K values in the matrices, based on the wing reference area S. Reference area is defined in the first matrix of the Aircraft Data File and is either explicitly given as the second item in the first column, or is calculated from a given W/S and the calculated takeoff gross weight. If you change S or W/S for a trade study, you must carefully consider whether C_{D0} and K must also be revised. For example, if you arbitrarily double the wing loading then the wing area (S) is halved, but the fuselage size is unchanged so C_{D0} would need to be recalculated.

However, if you are using the Aerodynamics and Weights Modules for data input, such trade studies can be performed simply by changing the appropriate inputs (S-ref, etc...) and rerunning the analysis to create a revised Aircraft Data File. RDS automatically fixes W and W/S when the analysis modules are used.

If you are analyzing an existing aircraft for which you have good drag polars (C_D vs. C_L), you can input those in the C_{D0} and K matrices then run a conversion routine. Enter the C_D's at $C_L = 0$ in the RDS C_{D0} matrix (for various velocities and altitudes), and in the K matrix, enter C_D for different lift coefficients and velocities at some representative altitude. Next, select the *Drag Polar Conversion* routine in the *Options* menu. C_{D0} and K matrices will be created for you.

Alternatively, you could enter the existing data as alternative drag polars as described above, then use Alternative Drag Polars for every item in both mission definition and performance calculation. This method is not recommended – it is clumsy and it is too easy to lose track of which drag polar to use.

If you wish to use the Oswald's Efficiency Factor method for rapid estimation of K, you should input 1 (one) when asked "how many lift coefficients?" and "how many velocities?", then enter 0 (zero) for the requested lift coefficient and velocity. When the matrix appears on the screen, enter your K value (1/πAe). RDS will understand that for this simplified method, K is not a function of C_L or velocity. However, the Leading Edge Suction method in the RDS Aerodynamics Module is far superior and almost as easy to use.

Engine data in the Aircraft Data File matrix must be installed' data including corrections for inlet recovery, inlet drags, nozzle drags, and bleed (for jets), or for propwash, tip Mach number, cooling drag, etc. (for props). This installed engine data can be calculated elsewhere and typed into the matrix, or can be developed in the Propulsion Module from uninstalled engine data.

It is not necessary to multiply the engine data by the number of engine, nor even to increase it if the engine is scaled up from the data available. During analysis, RDS will scale the engine data up to match either the input T/W (use this for "rubber" engines), or to match the thrust per engine times number of engines in the first Aircraft Data File array. In either case, the thrust is scaled to match the sea level static thrust value.

Since RDS scales all engine data to the T/W value provided, thrust trade studies can be done simply by changing the T/W value (and recalculating the weights). You cannot perform a thrust trade for a rubber-engine design by multiplying all thrust entries in the matrix by a constant, because RDS will simply scale them back to the previous values to produce the T/W or T you entered in the first matrix !

The Aircraft Data File includes thrust and specific fuel consumption for both afterburning and non-afterburning thrust settings. For non-afterburning jet aircraft, the afterburning matrices are left blank. Piston-propeller and turboprop engine data are stored in the same manner as for jet engines, namely, as net thrust and as thrust-specific specific fuel consumption, both versus speed and altitude.

For analysis of jet engines in the Propulsion Module, the manufacturer's uninstalled engine data is typed into RDS using the Aircraft Data File

Module. When saving this data, you should indicate that the matrix contains uninstalled engine data by using an .ENG extension for the filename. In other words, create a new file and enter the thrusts and C's, then save it giving a filename such as MYPLANE.ENG. Then you can go to the Propulsion Module and access it for installation corrections. The resulting installed engine data can then be stored into your Aircraft Data File (MYPLANE.DAT) when you run the propulsion analysis. Similarly, turboprop manufacturer's uninstalled data is entered and stored in a file with extension .TBP.

Uninstalled engine data can be obtained from the manufacturer, or may be approximated by using the engines in the back of Dr. Raymer's textbook and ratioing to the desired thrust level. Alternatively, RDS can use as input the uninstalled engine data tables produced by the jet engine design code ONX/OFFX, by Dr. Jack Mattingly, which is available from AIAA along with the textbook *Aircraft Engine Design* (AIAA Education Series, 2002). This permits you to design your own jet engine then install it in your aircraft for RDS performance analysis.

If you have engine data in fuel flow (lbs or kg per hour) rather than specific fuel consumption data, you may enter it in the C matrix then select the Fuel-Flow-to-SFC conversion routine from the options menu. You must have the thrust data entered as well before you run this routine. It will then convert your input values to C by dividing through by the appropriate thrust.

While piston-powered aircraft show little variation of specific fuel consumption with power setting, the same is not true for jets and turboprops. At a reduced power setting these turbine engines often consume far more fuel per unit thrust, in some cases more than double. There are several ways of dealing with this in RDS. The simplest option is to input typical "cruise" SFC's in the non-afterburning SFC matrix rather than inputting the max-thrust SFC's. This introduces some error if the thrust required during some mission segment is substantially less than or greater than the assumed cruise thrust settings.

A much better method, the RDS default, uses a semi-empirical equation developed by Dr. Mattingly, co-author of *Aircraft Engine Design*. This method needs no additional inputs and provides a realistic increase in the

SFC as thrust is reduced. You should check this equation's results versus your engine data to make sure that it is representative.

$$\frac{c}{c_{maxdry}} = \frac{0.1}{x} + \frac{0.24}{x^{0.8}} + 0.66 x^{0.8} + 0.1 M \left(\frac{1}{x} - x\right)$$

where x = T / Tmax dry

When using this method you should enter the SFC values for maximum dry (non-afterburning) throttle setting. This empirical equation can also be applied to turboprops, but with lesser accuracy.

Part power SFC adjustment is not applied to afterburning operation. If afterburning operation is selected for cruise or loiter conditions (by inputting 999 for thrust setting), RDS will calculate and apply the thrust setting required then apply the stored afterburning SFC values. This assumes that your jet engine has partial-AB capability. If not, you must ensure that your mission does not involve a cruise or loiter segment with afterburning selected (999) unless you are certain that the full afterburning thrust is actually required during that condition.

For a propeller-driven aircraft, engine data in the Aircraft Data File is also stored as thrust and thrust-specific SFC (lb/hr/lb-thrust or mg/Watt-s), just as it is for a jet aircraft. Calculation of propeller thrust and thrust-specific SFC is done in the Propulsion Module from inputs of horsepower, power-specific SFC, propeller efficiency, and other input data.

For piston-propeller aircraft the Aircraft Data File includes four propulsion matrices. Thrust and SFC at maximum power (full rich mixture) are stored in the first two engine arrays, normally used for takeoff and initial climb calculations. Thrust and SFC at economy mixture are stored in the second two engine arrays and are normally used for cruise, loiter, descent, and other conditions.

Turboprop aircraft have only one-each array of thrust and SFC. However, the Propulsion Module copies the same data into the "Max" thrust and SFC arrays so that you can run the same mission files for piston-prop and turboprop engines.

The Aircraft Data File Module has a powerful pair of menu options for data revision. ADD DELTA X and MULTIPLY X can be used to cause an input value to be added to or multiplied times the entire matrix or a specified row or column. This capability is very useful for 'fudge-factoring' for trade studies.

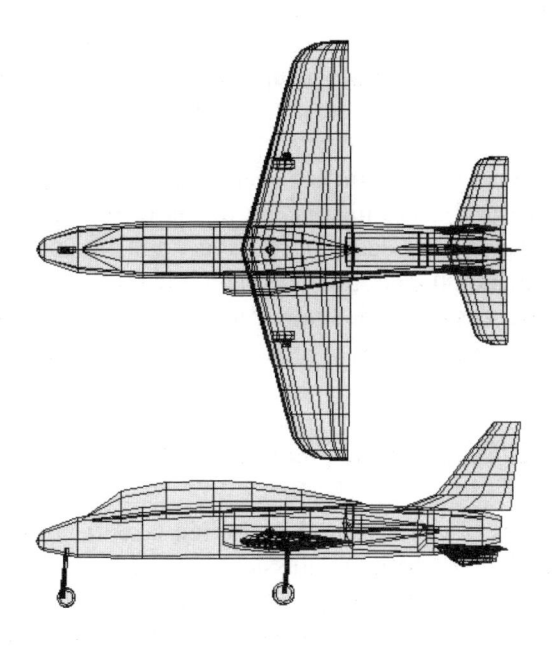

CAPABILITIES ANALYSIS MODULES

CAPABILITIES ANALYSIS OVERVIEW

Capabilities Analysis takes the results of the technical analysis (aerodynamics, weights, and propulsion) and determines air vehicle performance capabilities including range, speed, altitude, rate of climb, takeoff distance, and many other design requirements. This can also include the analytical process of "Sizing," which calculates the required takeoff gross weight of a new design to perform some given mission and range. These calculations are based upon the methods described in Dr. Raymer's textbook, *Aircraft Design: A Conceptual Approach.*

Cost analysis is also included in this section of the manual because the cost to attain a desired capability is the penultimate decider in aircraft development and purchasing decisions. Cost is normally done after Sizing, being based so much on the aircraft sized empty weight. Cost is also a key measure-of-merit in design trade studies, where the revised empty weight as calculated in the trade study is used to determine a revised cost.

SIZING & MISSION ANALYSIS MODULE

Sizing is the process of calculating the required takeoff gross weight of a new design to perform some given mission. Mission analysis determines the range capability of a design (new or existing) where the design takeoff gross weight is known and unchangeable. Both are done in the Sizing & Mission Analysis Module. Both rely on data in the Aircraft Data File, so it must be fully defined before the Sizing and Mission Analysis Module can be used.

A new mission profile is created by selecting the mission segments that form your mission. When you are done selecting mission segment types, RDS displays a matrix of your selected mission segments. You must then enter the required mission input values, such as cruise range, altitude, etc. When the mission is fully defined, select Run Sizing from the menu to execute sizing.

Figure 13 shows a typical summary sizing output. You can also select a full printout using lots of paper, or you can write the full sizing results to a text file.

```
    MISSION SEGMENT        MISSION SEGMENT WEIGHT     Wi/WO      FUEL BURN  (lbs-m)
                          FRACTION OR DROPPED WEIGHT             SEGMENT    TOTAL
 1  TAKEOFF SEGMENT                0.9584             0.9584      709.4      709.4
 2  CLIMB and/or ACCEL.            0.9736             0.9331      432.4     1141.8
 3  CRUISE SEGMENT                 0.9721             0.9071      444.0     1585.7
 4  CLIMB and/or ACCEL.            0.9950             0.9025       76.8     1662.5
 5  CRUISE SEGMENT                 0.9813             0.8856      288.6     1951.1
 6  KNOWN TIME FUEL BURN           0.9339             0.8271      999.3     2950.4
 7  WEIGHT DROP SEGMENT          400.0000             0.8036        0.0     2950.4
 8  CLIMB and/or ACCEL.            0.9800             0.7875      274.7     3225.1
 9  CRUISE SEGMENT                 0.9817             0.7731      246.2     3471.3
10  CLIMB and/or ACCEL.            1.0000             0.7731        0.0     3471.3
11  CRUISE SEGMENT                 0.9716             0.7511      375.0     3846.2
12  DESCENT SEGMENT                0.9900             0.7436      128.1     3974.3
13  LOITER SEGMENT                 0.9692             0.7207      391.0     4365.4
14  LANDING SEGMENT                0.9950             0.7171       61.5     4426.9
                                                        Reserve & trap :     265.6
                                                        Total fuel    :    4692.5

   Seg.  3   CRUISE :   487.2 kts at  45000.0 ft         RANGE     =    200.0 nmi
   Seg.  5   CRUISE :   806.4 kts at  35000.0 ft         RANGE     =     50.0 nmi
   Seg.  9   CRUISE :   806.4 kts at  35000.0 ft         RANGE     =     50.0 nmi
   Seg. 11   CRUISE :   487.2 kts at  45000.0 ft         RANGE     =    200.0 nmi
   Seg. 13   LOITER :   191.6 kts at    200.0 ft         ENDURANCE =      0.3 hrs

          TOTAL RANGE  =      500.0       TOTAL LOITER TIME  =       0.33
          FUEL WEIGHT  =     4693.0             EMPTY WEIGHT =   11258.2
   USEFUL LOAD  (-Wf) =     1110.0    AIRCRAFT GROSS WEIGHT  =   17061.2
```

Figure Thirteen. Typical Sizing Results

Velocities are entered as true airspeed or as Mach number. RDS checks to ensure that your input speed is neither below stall speed at that altitude, nor above maximum speed at that altitude and power setting.

To perform fixed takeoff weight mission analysis you must define a mission as above, then indicate one or more cruise or loiter mission segments as variable in length or duration. This is done by entering a one (1) in the indicated column position labeled *1=VaryRange* or *1=VaryTime*. When you select RUN SIZING, RDS will hold the aircraft takeoff weight constant and will vary the range or loiter time of all mission segments so indicated until the solution converges, then will print out the resulting range and loiter times.

The Sizing & Mission Analysis Module includes the use of a range credit for climbs and descents. In the case of climbs, the distance required to perform the climb is calculated, saved and subtracted from the next cruise segment.

Descent fuel burn and range credit depend largely upon pilot technique – the pilot can chop all power and dive down abruptly, or slightly reduce power and descend gradually over a long period of time. There are two methods of defining a Descent Mission Segment in RDS. Usually the simple method for defining a Descent Segment is adequate since the descent fuel consumption is normally small (mission segment weight fraction is nearly 1.0). The user just inputs a historical mission segment weight fraction and a range credit distance, which is subtracted from the requirement for the preceding cruise segment's range. A three (3) degree descent profile at cruise speed is commonly used, so the descent range credit can be estimated as the altitude lost divided by the tangent of three degrees.

The better method is labeled *Descent Analysis*, and requires inputs for throttle setting, descent speed, extra drag for speed brakes, etc. This method then calculates the descent rate, the range credit for descent, and the mission segment weight fraction.

One source of potential confusion in defining mission inputs is the use of external dropped stores, such as bombs for a fighter. These are "dropped" in the mission definition by providing a "Weight Drop" mission segment, indicating the weight dropped and the number of stores dropped. The weight dropped is the TOTAL weight dropped at that point in the mission, and is *not* multiplied by the number of stores you have input. When the sizing analysis reaches this point in the mission, the aircraft weight is decremented by this total weight. The Wo value in the Aircraft Data File must include this to-be-dropped weight.

The value you input for number of stores is used to calculate the extra drag of the external dropped stores before they are dropped. This number, times the D/q per store value you stored in the Aircraft Data File, is added to the parasite drag for all mission segments before the weight drop segment. In other words, droppable stores drag (D/q per store) is entered in the first array of the Aircraft Data File (.DAT), and the mission file merely indicates how many external stores are dropped along the mission.

Thus, to properly handle internal weapons being dropped you just need to decrement weight by the total weight dropped, using a Weight Drop mission segment weight fraction. Your takeoff weight must have included the weight of the stores being dropped. However, if those stores are carried externally you must do two additional things. First, you must define the additional drag per store in the column D/q per store in the Aircraft Data File, then you must include the number of stores being dropped in the Weight Drop mission segment along with the weight dropped.

There is a way to "trick" RDS which is sometimes useful for missions and trade studies involving external stores. When RDS encounters a value for the number of dropped stores and does NOT find stores D/q in the Aircraft Data File, it assumes a D/q of one. In other words, the value input as number of dropped stores is assumed to actually be the stores drag area (D/q) and is added to the aircraft drag on all mission segments before that Weight Drop mission segment. So, you can put actual stores drag area (D/q) directly in the mission file in the "number of dropped stores" entry, and not bother putting it in the Aircraft Data File. Note that the number of external stores does not have to be an integer value.

Two caveats must be noted if you are using this "trick" – first, this assumes that stores D/q does not change with Mach number. If you want to input drag variation with Mach number you must do it the "real" way, by entering a D/q table in the Aircraft Data File. Another caution – if you later decide to include stores D/q in the Aircraft Data File, RDS will multiply those values times the number in the Weight Drop mission segment to calculate total D/q. You must change the method in both places.

While confusing at first, this convoluted method gives great flexibility for dealing with all sorts of external and internal stores carriage with a minimum number of additional inputs.

In some rare cases involving "out-and-back" missions, the climb and descent range credits can cause problems. For example, RDS was used to calculate the range of an L-1011 launching the Orbital Sciences Pegasus booster. This mission involves a steep climb to launch the booster, just before beginning the return to base. RDS tried to apply the range credit for the climb to the return range! In such a case, a "Weight Drop" mission segment can be used to indicate the turn-around point, canceling all range credits prior to the turn-around point. This can also be used for the case where, for some reason, the customer requirements specify that no range credits may be taken.

Alternative Drag Polars in the Aircraft Data File can be used for mission or performance calculations when the drag polars determined from the stored C_{D0} and K would not reflect the aircraft configuration. This could be, for example, when an airliner has flaps deployed for climb or descent, or when an aircraft with a variable sweep wing has it swept at an angle other than the assumed baseline. Up to seven alternative drag polars can be used, and are entered and stored as paired columns of Cl and CD in the first Aircraft Data File matrix.

To use alternative drag polars which have been stored in the Aircraft Data File, enter the drag polar number (1-7) at the item "ALT POLAR #" in the Performance matrix or in the Mission file in Climb, Cruise, Loiter, or Descent Analysis segments.

Sizing Sensitivity tells the sizing routine when the calculated gross weight is close enough to the "guess" gross weight, and can be changed in the options menu. Using a value of, say, .001 rather than the default of .0001 will speed up execution especially for trade studies, and may also avoid the rare occurrence of an infinite loop in sizing analysis. However, the sizing results may be slightly off because RDS will accept convergence when the calculated W_0 is still significantly different from the "guess" value of W_0.

Another option in the Sizing & Mission Analysis Module gives the ability to add or delete mission segments. This is selected from the menu options at the bottom of the matrix. The current mission segment types are listed, with the first item highlighted. Select the mission segment which you wish to delete or follow with a new mission segment. Press Enter or the left mouse button then input A to add a new segment, or D to delete the indicated segment. If you press A, a list of mission segment types will appear. Select the desired type. When finished adding mission segments you must enter all data required for the new segments back in the main mission matrix.

This ADD/DEL option can also be used to list out all mission segments in your mission model for review. Then you can cancel it without changing anything.

The Options menu permits changing the definition of Service Ceiling and the maximum number of sizing iterations along with other options.

Trade studies can be performed in several ways. For mission and performance trades such as cruise range or takeoff distance, simply change the appropriate input value and rerun the analysis. Write down the answers and use them to make a graph of the changed input value versus the results, such as range versus takeoff gross weight (Excel^{TM} is a great tool for the graphing).

For trade studies involving aircraft characteristics such as SFC or parasitic drag, trades can be done simply by modifying the Aircraft Data File. Each data matrix such as C_{D0} or thrust can be modified by addition or multiplication of matrix data by an input constant (one of the commands at the bottom of each array in the Aircraft Data File). You can apply this to the entire array, or for a more sophisticated trade study, apply a constant to individual rows, columns, or data items. For example, you could use this method to do a drag trade study of an arbitrary increase in supersonic C_{D0} without a change in subsonic C_{D0}.

Trade studies can also be done by making changes in the inputs of the Aerodynamics, Weights, and Propulsion analysis Modules, running *Do Analysis* to create a new Aircraft Data File, and using those results to rerun Sizing and Performance calculations.

PERFORMANCE ANALYSIS MODULE

RDS calculates all of the commonly-used performance parameters such as rate of climb and takeoff distance, and produces a variety of performance graphs. Performance analysis requires that the Aircraft Data File be fully defined as described above. When Performance Analysis is selected, a menu of options appears including *Flight Envelope, Range Optimization, Rate of Climb, fs Calculation, Ps and Turn Rate,* and *Takeoff, Landing, Accel, Ps, Turn, & Climb*. The first five are obvious, and will lead you through the required inputs to do the indicated analysis. Figure 15 shows the Rate of Climb graphical output.

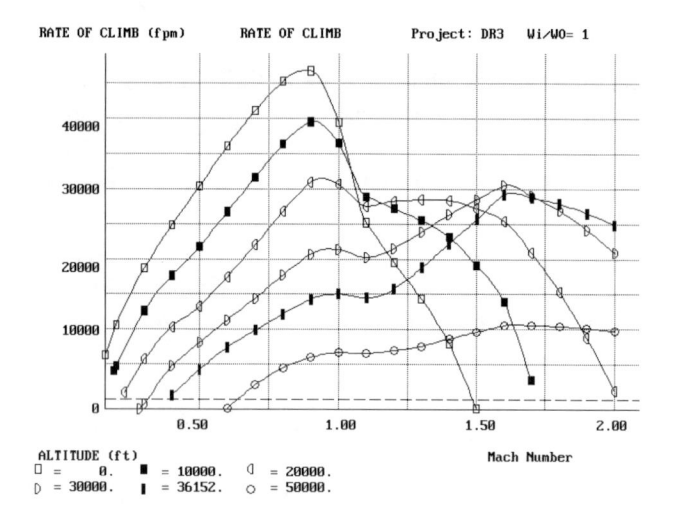

Figure Fifteen. Rate of Climb

The ratio Wi/W0 is the aircraft weight at which you wish performance to be calculated, divided by the aircraft takeoff weight. Performance may be calculated at takeoff weight (Wi/Wo=1), or at some arbitrary weight ratio such as 0.8. You can use the listing of Wi/Wo from the sizing analysis results to find the actual weight ratio at some point during the mission, such as at the beginning of combat. You can also input an actual aircraft weight at which you wish performance calculations to be done, letting RDS divide by takeoff gross weight to get the ratio.

The menu selection *Takeoff, Landing, Accel, Ps, Turn, & Climb* permits you to define and store the inputs for those analysis items, entered in an array of columns, one for each type of performance to be calculated. First select the desired performance calculations from the list of options which appears. Note that the Ps and Turn Rate Option is used to calculate those values at one specific altitude and velocity, unlike the similar option on the Performance menu in which Ps and Turn Rate values are calculated for the entire flight envelope.

When you are done selecting performance types, a matrix of the required input values will appear including normal default values of many parameters. Enter the required values, then select RUN PERF from the menu. Inputs for gear drag as well as an option for engine-out allow performance calculations for engine-out go-around and similar requirements.

The performance input matrix has, at the bottom, entries for Required Values and for Requirement Codes. These are printed with the performance analysis results for comparison of goal to actual. They are used in the RDS-Professional Carpet Plot and Optimization Module as the optimization constraints.

The Options menu in Performance allows setting various mouse and graphing options. It also includes the ability to specify a descent angle for landing approach (three degrees is the default, typical of transport aircraft).

COST ANALYSIS MODULE

Development and procurement cost analysis is done using a modified DAPCA-IV method[*]. Inputs are via a spreadsheet-like matrix, similar to the other RDS analysis modules. All cost entries must be completed. To calculated costs in other than 1998 dollars it is necessary to enter an economic escalation (inflation) factor, defined as "then-year-dollars"/1986 dollars. Also, hourly rates for engineering, production, etc. should be entered in then-year dollars.

DAPCA tends to underestimate costs for new-generation fighter aircraft, so a "DAPCA Fudge Factor" of 1.25 is recommended. DAPCA seems to over predict transport aircraft costs so a fudge factor of 0.9 is recommended. DAPCA was not developed with general aviation aircraft in mind, but some users report that DAPCA with an overall fudge factor of 0.25 gives approximate results. This is very crude, and general aviation cost estimates should be calibrated by use of actual data for a similar aircraft to create a DAPCA Fudge Factor suitable for your design.

---

[*] Development and Procurement Cost of Aircraft – see Raymer's textbook

For an aircraft with composite materials, a "Materials Fudge Factor" of 1.5 is suggested.

An "Investment Cost Factor" of 15% is typical. This includes the contractor's cost of borrowing or using money as well as contractor profit. See Dr. Raymer's textbook for more information on cost analysis using the DAPCA model.

There are two ways to define engine cost in this module. If engine cost is known, enter it directly in the appropriate place as cost per engine. If engine cost is unknown, RDS will estimate it from inputs for thrust, maximum Mach number, and turbine inlet temperature. If this estimate is being used, leave engine cost blank to tell RDS to calculate it using the inputs. For a piston-propeller aircraft, you must enter engine cost – RDS has no estimating method.

Life Cycle Cost and Airline Economic analysis are calculated by yearly accumulation of costs based upon user inputs, adjusted for the inflation rate. These inputs are defined in the input matrix based upon concepts in the textbook.